FOR LOVE OF REGIMENT

Also by Charles Messenger:

For Love of Regiment Volume I Leo Cooper
Trench Fighting 1914–18 Pan-Ballantine
The Art of Blitzkrieg Ian Allan
The Observer's Book of Tanks Frederick Warne
Terriers in the Trench: The Post Office Rifles at War Picton Publishing
The Unknown Alamein Ian Allan
Cologne: The First 1,000 Bomber Raid Ian Allan
The Tunisian Campaign Ian Allan
The New Observer's Book of Tanks Warne/Penguin
Bomber Harris and the Strategic Bombing Offensive 1939–45
Arms and Armour
Armies of World War 3 Bison/Hamlyn
Modern Combat Weapons: Helicopters Franklin Watts
Tanks Kola Books
Modern Combat Weapons: Combat Aircraft Franklin Watts
The Commandos 1940–1946 William Kimber/Grafton
Anti-Armour Warfare Ian Allan
Northern Ireland: The Troubles Bison/Hamlyn
The Steadfast Gurkha: 6th Queen Elizabeth's Own Gurkha Rifles 1948–1982
Leo Cooper
A History of the British Army WH Smith/Bison Group
The Second World War Franklin Watts
The Middle East Franklin Watts
Hitler's Gladiator Brassey's
Middle East Commandos William Kimber
World War Two Chronological Atlas Bloomsbury
Battle of Britain WH Smith
World War II in the Atlantic Bison Group
The Last Prussian Brassey's
Great Military Disasters Bison Group
The Century of Warfare HarperCollins

FOR LOVE
OF REGIMENT

A History of the British Infantry
VOLUME TWO
1915–1994

by
CHARLES MESSENGER

LEO COOPER
LONDON

First published in Great Britain in 1996 by
LEO COOPER
190 Shaftesbury Avenue, London WC2H 8JL
an imprint of
Pen & Sword Books Ltd,
47 Church Street, Barnsley, South Yorkshire S70 2AS

© Charles Messenger, 1996

A CIP record for this book is available
from the British Library

ISBN 0 85052 422 9

Typeset by CentraCet, Cambridge
in Linotron 10pt Plantin

Printed by Redwood Books Ltd.
Trowbridge, Wilts.

This book is dedicated to the British Infantryman,
past, present, and future,
especially Major Gregory Blaxland, late The Buffs
1918–1986

Contents

INTRODUCTION

VOLUME One of *For Love of Regiment* ended with the British Expeditionary Force in France about to draw breath after its exhausting struggle to hold back the Germans at Ypres during the late autumn of 1914. The sequel picks up the story from that point and brings the history of the British Infantry up to the present day.

If anything the challenges facing the Infantry and, indeed, the British Army as a whole have been ever greater during the past eighty years than in the previous 250 years. Certainly, the variety of tasks that the Infantry has been called upon to perform has been increasingly varied. Overshadowing all has been the two world wars, but interwoven with these has been the British infantryman's traditional role of defence of empire, and post-1945 acting as the rearguard during withdrawal from it. In recent years, too, there have been the long agony of Northern Ireland, service with the United Nations, and the Cold War. There has, too, been the unexpected; the campaign amid the desolate terrain of the Falkland Islands in 1982 and the armoured blitzkrieg in the sands of Kuwait in 1991 are but two samples.

The past fifty years have also witnessed a growing uncertainty over the survival of the regimental system, for so long the Infantry's bedrock. Time and again it has been under threat as circumstances have forced the British Army to reduce its strength. Yet a study of Appendix One will reveal that the Infantry has endured drastic expansions and cuts ever since the British Army, as we know it today, was created by King Charles II in 1660. At present the Infantry is in the midst of a steady contraction, which may well continue if the much hoped for permanent peaceful resolution to Northern Ireland's affairs is found. To compare the post-Options for Change list of Foot Guards and Line regiments as it exists today at Appendix Two with that of August, 1914 (Appendix, Volume 1) makes sober reading, but it might well become slimmer in the near future.

Yet the regimental system still survives and I hope that the arguments in support of it that I advance in this volume will ensure that it is maintained for the foreseeable future. Indeed, we approach the second millenium AD in a world that is more uncertain than we who are alive today have ever known. It is all too easy to advocate change for the sake of change, but when all around us is fluid we need constants or otherwise we will lose our way. The regimental system, in spite of its flaws, has served us well in the past. There is no reason why it should not continue to do so, especially in view of the challenges, many as yet not envisaged, with which the British Army will inevitably be faced in the future.

CHARLES MESSENGER
London
November 1994

'They marched back from the battle [Cassino, Italy, 1944] in the way of the infantry, their feet scarcely leaving the ground, their bodies rocking mechanically from side to side as if it was the only way they could lift their legs. You could see that it required the last ounce of their mental and physical courage to move their legs at all. Yet they looked as though they could keep on moving like that for ever.

Fred Majdalany *The Monastery*

CHAPTER ONE

France and Flanders
1915–1918

THE failure of the Germans to break through at Ypres marked the end of active warfare on the Western Front for 1914, apart from two hastily mounted and disastrous attacks by the French in December and one or two minor skirmishes on the British front. It was as well for the British Expeditionary Force (BEF) that there was such a lull. Its infantry especially were exhausted by the fighting of the past four months and their battalions were, in many cases, now skeletons. They needed the chance to recoup their strength and to grapple with the complexities of trench warfare.

By January, 1915, the BEF had grown to a strength of eight British and two Indian infantry and three British and two Indian cavalry divisions. It occupied a sector 28 miles long, running from St Eloi, just south of Ypres, southward to La Bassée. They were, however, in an uncomfortable position. Many of the hastily constructed trench systems were based on drainage ditches, part of the intricate system of controlling surface water in Flanders. This had now been upset and matters were made worse by heavy rain. Consequently, flooding became widespread. Matters were not helped by the fact that the Germans tended to occupy the higher ground and possessed seemingly unlimited artillery and mortars. In contrast, the British had all but exhausted their stocks of artillery ammunition and the guns were strictly rationed. They also had no answer to the mortar, called the *Minenwerfer* by the Germans and which had been part of their siege train to reduce the Belgian forts in August, 1914.

Bruce Bairnsfather, a subaltern in the Royal Warwicks and creator of the cartoon characters 'Ole Bill' and 'Young Bert', who so epitomized the character of the BEF during the first part of 1915, described conditions as follows:

'Select a flat ten-acre ploughed field, so sited that all the surface water of the surrounding country drains into it. Now cut a zig-zag slot about four feet deep and three feet wide diagonally across, dam off as much water as you can so as to leave about a hundred yards of squelchy

1

mud; delve out a hole at one side of the slot, then endeavour to live there for a month on bully beef and damp biscuits, whilst a friend has instructions to fire at you with a Winchester every time you put your head above the surface.'

Thus much effort had to be put into keeping the trenches habitable, but much else had to be done besides. Private Frank Richards of the 2nd Royal Welsh Fusiliers:

'At night, we numbered off, one, two, three, one, two, three – ones up on sentry, twos and threes working. Every evening at twilight the order would come "Stand to!" and every man in the trench would get up on the fire-step and gaze across no-man's-land at the enemy's trench. The same thing would happen at dawn in the morning. After standing to about five minutes the order would come "Stand down!" A sentry would be on from stand to in the evening until stand to next morning, which during the long winter nights meant fourteen or fifteen hours continual standing on the fire-step and staring out at no-man's-land. At night all sentries stood head and shoulders above the parapet; they could see better and were less liable to be surprised . . ., Twos and threes were working all night, some carrying R.E. material from Houplines – this consited of duckboards for laying on the bottom of the trench when the water was cleared out, barbed wire, sandbags and other material for building trenches. Some were carrying rations, others filling sandbags. . . . Some were putting out barbed wire in front, and others were strengthening the parapet. . . . During the day we were working in reliefs, and we would snatch an hour's sleep, when we could, on a wet and muddy fire-step, wet through to the skin ourselves.'

Another private soldier, a Territorial in the 5th Londons (London Rifle Brigade), who had been in France for a bare three weeks, noted in his diary on 27 November, 1914: 'Everything and everybody plastered with mud; mud on your hands, and down your neck and in your food, and bits of mud in your tea.' Yet, a few days later, returning to the front line after two days' rest, to relieve a Regular battalion, he noted that it was a sign that 'the Staff trusts us' and that he was 'proud to work with and relieve these splendid Tommies; most of them reservists'. Indeed, this was the attitude prevalent among the Territorial battalions sent to France in 1914. They had the deepest respect for the Regulars and were almost pathetically keen to learn from them. The Territorials, too, wanted to disprove War Minister

Kitchener's view that they were useless as soldiers, a view that had quickly permeated the Regular suvivors of the 1914 BEF.

In the cold and muddy conditions that existed in Flanders during the winter of 1914–15 a high sickness rate might have been expected, but this was not generally so. The cold was kept at bay by keeping active. Thus the routine described above by Frank Richards might seem very tough, but it prevented hypothermia. The main result of damp, though, was trench foot, the symptoms being the foot turning red or blue and inflicted with chilblains, with, in extreme cases, gangrene setting in. The way of overcoming it was frequent changes of socks and the application of grease or whale oil. To ensure that the troops did this meant frequent foot inspections by the company officers and the incidence of trench foot became a means of gauging the quality of a battalion; good ones did not suffer from it. Interestingly, there were a number of cases of trench foot in the Falklands in 1982 and the lessons learnt during winter 1914–15 had to be hurriedly relearnt.

There were also other ways in which the morale and efficiency of a battalion could be gauged besides the length of its sick list. One was the state of its trenches. The good battalion was constantly working to improve its defences, sending out wiring parties each night to repair and strengthen the barbed wire in front of the trenches. Damage caused by enemy fire was repaired immediately and strict hygiene discipline was rigorously enforced, especially over the latrines, which were usually dug in the rear of each line of trenches. The bad battalion made little effort, handing over its trenches in a filthy state.

Fire discipline was another indicator. At night there would often be bursts of fire, usually caused by a nervous sentry thinking that the enemy were approaching or because he was bored. This would be taken up by others and the enemy would reply in kind. Known as 'wind up', well disciplined battalions seldom indulged in it and it was usually those who were new to the trenches who were to blame. There was also the conflict between the policy of 'live and let live' and those battalions who considered it their duty to dominate no-man's-land. While there were unofficially recognized quiet and active trench sectors, usually because of the nature of terrain, aggressive battalions would take little notice of this, and often made themselves unpopular with others who relieved them in sectors which had previously been quiet but had now become unpleasantly active.

A major problem in the line battalions in France by the end of 1914 was a severe shortage of junior officers. Casualties among these had been especially heavy and most junior Regular officers still at home had been sent to the New Armies. One solution resorted to in France was the commissioning of warrant officers and senior NCOs. RSM Murphy of the

2nd Royal Welsh Fusiliers was one of these. 'There was I, a thousand men at my control, the Commanding Officer was my personal friend, the Adjutant consulted me, the Subalterns feared me, and now I am only a bum-wart and have to hold my tongue in the Mess.' Nevertheless, those commissioned from the Warrant Officers' and Sergeants' Mess of the 2nd Royal Welsh Fusiliers more than justified their selection. One, W.H. Stanway, who was a Company Sergeant Major in 1914, was commanding a battalion of the Cheshires before the end of 1917 and finished the war with a DSO, OBE and MC.

Another source of officers which was exploited at the beginning of 1915 were the Territorial battalions, who had many of officer potential in their ranks. Indeed, during the height of First Ypres Sir John French personally inspected the newly arrived 28th London Regiment (Artists Rifles), picked fifty men out of the ranks and had them commissioned Second Lieutenants on the spot. Next day many of them found themselves in action in command of Regular troops, wearing their private's uniforms with the addition of stars on their shoulder straps. Some were killed before they could be gazetted. Sir John then ordered the Artists Rifles to Bailleul, where his GHQ was situated, in order that they should take on the permanent responsibility for producing officers. Here they ran a four-week course, which included drill, map reading, machine-gun work, trench fighting, billeting, field engineering and elementary surveying. Successful students were then given a week's leave at home in order to kit themselves out as officers and then joined their battalions at the front. It took, however, a few weeks to get these courses organized and in the meantime those selected for commissions merely arrived at their new battalions in their private's uniforms and were then sent away for a week to equip themselves. They had to learn to be platoon commanders the hard way, on the job, and those who failed to meet the exacting standards expected of them were returned from whence they came.

The Inns of Court Regiment, whose drill hall at Lincoln's Inn in London emphasized the fact that it drew largely on the legal profession for its members, and the universities also ran officer training courses. Early in 1916, however, it was decided to put officer training on a more formal basis. To this end, officer cadet battalions were set up in Britain and those recommended for commissions were sent on a four-month course at these.

Another means of easing the shortfall of young officers was to attach officers of a different cap badge to battalions that were very short. Thus Robert Graves, who had joined the Royal Welsh Fusiliers Special Reserve in 1914, initially found himself serving with the 2nd Welch Regiment when he first went to France in May, 1915. Both it and the 1st Battalion had suffered especially heavy casualties and could not make good their losses from their own resources. Indeed, Graves noted that on joining the 2nd Welch there were only three company officers who wore its cap badge and that even the quartermaster came from another regiment. Because the Battalion was so desperate for officers Graves was made much welcome. In contrast, when, a few months later, he was posted to the 2nd Battalion of his own regiment, an officer of the East Surreys attached to it was disparagingly referred to as 'the Surrey man'.

Many barely trained recruits were also finding themselves joining Regular battalions at the front. Percy Croney joined the 12th Essex Regiment, a Kitchener battalion, at the beginning of 1915. That summer his battalion was broken up to provide drafts for the two line battalions, the 2nd in France and the 1st in the Dardanelles. He found himself posted to the latter and on arrival at Gallipoli he was immediately ordered to relieve a sentry in the front line. Almost the first words that the sentry said to him were: 'When a recruit speaks to a man with seven years' service and more, he must address him as "soldier", short for "trained soldier", and there are thirteen years' service behind me.' Indeed, the Regular line battalions tried their hardest to maintain their prewar attitude to life. The 1st and 2nd Royal Welsh Fusiliers even played polo during the first two years of the war in France and Flanders, and the latter subjected all its junior officers to the rigours of the Battalion Riding School when out of the trenches. It was only the devastating casualties of the 1916 Somme battles that brought about the end of all peacetime habits in some Regular battalions.

Later in the war less and less note was taken of cap badges, especially after a period of heavy casualties, and men often found themselves posted to any battalion of any regiment. Thus, Croney, who was wounded at Gallipoli, recovered in time to accompany the 1st Essex to France from the Middle East, only to be wounded again. Reporting to 3rd (Depot) Battalion The Essex Regiment, having been medically downgraded was bad enough, since the atmosphere was 'all grumbling' and lacked the purposeness of a line battalion, but worse was to follow. He found himself posted to the 2nd/5th Scottish Rifles, a home service battalion. 'Such a battalion of our own Country Regiment would be bad enough', but even the King would never transfer a man from the regiment in which he took the 'King's Shilling', yet some filthy civilian in the War Office, or some politician, has taken upon himself to do so.' Eventually he was posted to the 2nd Cameronians in France towards the end of the Third Battle of Ypres and seems to have settled down well enough, although he was to be wounded a third time and captured during the German March, 1918, offensive.

Certainly, though, there was a good deal of insensitivity over replacement drafts. Captain Hitchcock of the 2nd Leinsters noted in his diary in December, 1916, that two squadrons of the 5th Royal Irish Lancers were being disbanded and it was hoped that they would join the Leinsters. Instead they were sent to the 13th Middlesex and the Leinsters received a draft of Dorsets. 'This was an injustice to us, and to the Lancers.'

Individuals, though, sometimes took drastic steps to ensure that they were posted back to their own battalions after being wounded or invalided home. Charles Carrington, whose *A Subaltern's War* (written under the pseudonym Charles Edmonds) is one of the classics of trench literature, had served with the 1st/5th Royal Warwicks in France, winning the MC as a company commander during Third Ypres, but had then been sent home on leave at the beginning of 1918, during which time he fell sick. Posted, on recovery, to the 3rd/5th, a depot battalion, he became concerned that he might be posted to an active battalion other than his 'beloved First-Fifth', which was now in Italy. He pulled strings with the War Office and eventually got his way, but it was not until late October, 1918, that he was finally ordered to Italy. In the meantime he was congratulated by his brother officers 'on having invented a new and ingenious way of "dodging the column"' Robert Graves, on the other hand, tells of a draft of 1st Royal Welsh Fusilier veterans who deliberately overstayed their leave in order to avoid being sent to Mesopotamia. They were placed under arrest, and were quite happy to be sent back to their own battalion in France in handcuffs rather than go to another that they did not know.

The beginning of Spring, 1915, saw the first of the Territorial divisions arrive in France. Until then the only Territorial division sent abroad had been the 42nd (East Lancashire), which had been sent to Egypt in September, 1914, to relieve Regular troops on garrison duty in Egypt, and three others sent to India for the same purpose. The first to arrive in France was the 46th South Midland in February, 1915, and this was to be followed during the next few months by a further six divisions. These were first blooded at Aubers Ridge and Festubert.

That summer of 1915 the first of Kitchener's New Army divisions, his alternative to relying on the Territorial Force, began to arrive in France, beginning with the 9th Scottish and 12th Eastern in May.

The first major battle in which all three types of division – Regular, Territorial, New Army – took part was Loos in September, 1915. This was part of Joffre's efforts to bite off the huge German salient bounded by the Champagne in the south and Arras in the north by attacks aimed at double envelopment of it. The initial British assault, which was part of the blow in the north, was to be mounted by six divisions attacking on a six-mile front. The brunt was borne by the Regulars of 1st Division, the 15th Scottish (a K2 or second wave Kitchener division), and the London Territorials of the 47th Division. H-hour was 0630 hours, and the attack was accompanied by gas, the first time that the British had used it. However, on release, much of the gas hung about in no-man's-land or drifted back into the assault trenches. On the extreme right the Londoners, using dummies manipulated by strings to draw the enemy's fire, secured all their objectives within three hours, with the 18th Londons (London Irish) kicking a football ahead of them as they advanced. This was achieved with incredible coolness. One eyewitness:

'The air was vicious with bullets. Ahead the clouds of smoke, sluggish low-lying fog, and fumes of bursting shells, thick in volume, receded towards the German trenches, and formed a striking background for the soldiers who were marching up a low slope towards the enemy's parapet, which the smoke still hid from view. There was no haste in the forward move, every step was taken with regimental precision, and twice on the way across the Irish boys halted for a moment to correct their alignment.'

The 15th Scottish were, by contrast, less measured in their approach, but just as successful. Prior to the assault Piper Laidlaw of the 7th King's Own Scottish Borderers encouraged his comrades by marching up and down the parapet under heavy fire playing 'Blue Bonnets O'er the Border', the regimental march, until he fell wounded, a performance which won for him the Victoria Cross. The Scots then speedily overran Loos and Hill 70. In the words of Lieutenant Turnbull of the 8th Seaforths, who started the battle twenty-one officers and 755 other ranks strong and came out of it just a few hours later with just two officers and thirty-five men standing: 'Our men, deprived of nearly all their officers, took upon themselves practically a holiday mood, and moved about Hill 70 and the village of Loos looking for their friends.' When they tried to push forward beyond Hill 70 they came under heavy fire from as yet untouched fresh German trenches and were forced to pull back. The 1st Division also managed to break through at Hulluch, albeit with heavy casualties, and looked poised for a decisive breakthrough. Unfortunately, as so often happened in attacks across trenches, the reserves were not deployed in time and the survivors of the attack were too weak to resist the inevitable and quickly mounted German counter-attack.

Even so, success would still have been possible if the main attack reserve of three divisions had been immediately to hand. One of these was the newly formed Guards Division. The original BEF had included two Guards brigades, 1st and 4th, but only the latter was 'pure', with four Guards battalions, while the former consisted of the 1st Coldstream and 1st Scots Guards, and two line battalions. As further Guards battalions arrived in France, the decision was taken to create a dedicated Guards division, which came into being in August, 1915. As such it was to find itself used as a 'fire brigade' to overcome crises. Included in the Division was a new Guards regiment, the Welsh Guards. This had been formed in February, 1915, at the instigation of King George V. Originally the Royal Welsh Fusiliers, as the senior regiment of Wales, were invited to become Guards, but declined on the grounds that they would lose the character and traditions that the Regiment had built up over more than 200 years. Consequently the new regiment drew its men from Welshmen already serving in the Brigade of Guards, and its first duty was to mount the King's Guard at Buckingham Palace on St David's Day, 1 March, 1915.

The other two divisions in reserve were the recently arrived New Army 21st and 24th. French had insisted that he retain control of these reserves, rather than hand them over to Haig, whose First Army was conducting the battle, and the result was that they were positioned too far back and only began to move up at midday on the 25th. That evening saw them still on the march and Haig decided to use the 21st and 24th to spearhead a renewed attack early on the following day. Bewildered, tired, hungry and footsore, with no opportunity to view the ground over which they were to attack, these New Army men were pitchforked against now strongly reinforced German positions. Advancing across open country in, as one German war diary put it, 'ten ranks of extending line . . ., each one estimated at more than 1,000 men, and offering a target as had never been seen before, or even thought possible,' they were a perfect target for artillery and machine guns. Even so they kept advancing and eventually it was only the German wire that foiled them. In just three and a half hours they suffered over 8,000 casualties and the survivors streamed to the rear, totally numbed by what they had experienced. As they passed the head-quarters of their Corps Commander, General Haking, he asked them what had gone wrong. 'We did not know what it was like,' they answered. 'We will do it alright next time.' That encapsulated the spirit of Kitchener's men.

It was left to the Guards Division to restore the situation. They attacked and retook Hill 70 on the 27th. One who watched the attack called it 'the most wonderful sight of my life the Guards moving in open country under a curtain of fire. They marched, as if in Hyde Park, and I was very proud to think that some of them were friends of mine.' Certainly their success gave an enormous boost to morale, but the chance of breaking through had been lost and the battle gradually petered out, costing French his com-mand. Recognizing the inexperience of the New Army divisions, steps were taken after the battle to stiffen each by exchanging one or more battalions with Regular ones.

The winter of 1915–16 saw trench warfare settle down into a recognizable pattern. Corps headquarters were generally static, each being responsible for a particular sector. Divisions spent some three months on average under command of a particular Corps HQ before being withdrawn for rest and then deployed elsewhere. Each division usually had two brigades in the line and one in reserve, with the former normally having two battalions in the trenches and two in immediate reserve. In turn, the forward battalions would have two companies manning the front line and two in the support line. The normal trench tour would last 7–10 days. The routine in the trenches remained much as described earlier in this chapter, with night generally being turned into day and vice versa. Besides improving and maintaining the trenches themselves, there were more aggressive activities. Patrols, usually consisting of an officer, NCO, and perhaps one or two men, were sent out to check on the enemy's wire and to detect what he was up to in his trenches. Sometimes there would be clashes with German patrols bent on the same purpose. The trickiest part was often returning to one's own lines and numerous casualties were caused by jumpy sentries, as often as not not properly briefed on the patrol. Sometimes there would be a demand from above for a raid to be mounted in order to secure a prisoner for intelligence purposes. These were nerve-wracking operations. The usual tactic was to use artillery to seal off a section of the German line and then send a party dashing across no-man's-land and into it. A brief scuffle with the defenders and then, if they were lucky, a return dash, dragging their prisoners with them, in the midst of machine-gun fire, mortars and artillery from the now thoroughly alerted Germans. Often the aftermath was further patrols into no-man's-land to locate and bring back the wounded. Even during the very quiet periods there was still a steady trickle of casualties. Artillery and mortars were the main cause, although often the Germans would be very regular in their timings of their daily 'hate', as it was called. Snipers, too, were often very active and extremely accurate. Advances in medicine, especially the discovery of penicillin and better surgery techniques, meant, however, that a wounded man had a significantly better chance of survival than ever before. During the war on the Western Front only 7 per cent of casualties died at battalion aid posts, and 16 per cent at the casualty clearing stations, the next port of call for the wounded man and where the casualties were categorized according to the degree of surgery required, with those for whom there was little hope being placed in the 'moribund ward'. At base hospitals, on the other hand, 6 per cent died. Most soldiers dreamed of getting a 'Blighty' wound, not too incapacitating, but one that would get them back to Britain for a long spell.

There was also an improvement in the food. In the trenches themselves the troops subsisted on tinned rations, the two staples being corned beef, known as 'bully', and Maconachie, a meat and vegetable stew produced by a manufacturer of that name. They were also supplied with bacon, cheese and jam. Plum and apple was the most usual variety of the last-named, and also gave Plumer, whose Second Army defended the Ypres Salient for so long, his affectionate nickname. Later in the war, when food became generally scarcer thanks to the U-boat campaign, this gave way to rhubarb and marrow, which was not popular. Bread, too, was normally available, baked by the Army Service Corps. When it was not the troops had to make do with army biscuit, and carried an 'iron ration' of these and bully with them at all times. There was, too, the Tommy's staple, tea. The soldiers brewed this on their individual Tommy cookers, which used solid fuel, putting the tea leaves in the water before it was boiled and then adding condensed milk and sugar. It had to be strong to disguise the taste of petrol from the cans in which water was so often brought up to the trenches. While the individual soldier usually cooked for himself, or with his 'chum', in the trenches, the good battalion quartermaster would sometimes arrange for hot stew in thermos containers to be brought up if his men had been involved in particularly heavy fighting. Away from the trenches cooking was centralized by companies, even when the battalion was on the march from one sector to another. It was a common sight to see the horse-drawn cookers with the cooks stirring the stew as they marched so that it would be ready to serve at the midday halt. The only alcohol the soldier was allowed in the trenches was his daily tot of rum, usually taken in his tea. One or two divisional commanders, notably Pinney of the 33rd Division, forbade even this, fearful of encouraging the British soldier's traditional weakness. Out at rest, though, the soldier could, within reason, visit the local *estaminet* and drink wine or watery beer and order his favourite egg and chips. Sometimes, too, especially at Christmas, national and regimental days, the officers, at times with the help of the regimental Comforts Fund, would produce barrels of beer and special menus for their men.

The officers themselves generally messed by companies, even in billets, as they had done in the Peninsula a hundred years earlier. Before going into the trenches delicacies were purchased for the company 'mess box' in order to supplement the basic ration, although much depended on the company cook on how well his officers fared, as R.C. Sherriff made plain in his classic trench play *Journey's End*. Because of the number of visitors to the trenches, battalion HQ and company messes often used to go to great lengths to be hospitable, considering this a reflection on the reputation of the battalion, offering a wide range of drinks, although whisky, drunk with chlorinated water, was always the staple, and elaborate meals. There was, however, another reason for this. Brigadier General Kentish, who ran the Senior Officers' Course*, which was primarily for the benefit of potential battalion commanders, at Aldershot, gave a special lecture on running an officers' mess in the field. The time is August, 1917:

'I have personally lived in many Messes, and have by design and on principle had many a meal in many other Messes . . . and I definitely state that in those Messes in which the Senior Officer had that peculiar stomach and those ideas of living [economizing] by making his officers feed merely off the basic ration, and those ideas of living, a general air of depression has been manifest throughout the meal. Listen to this:- Bully-Beef, Biscuit, Cheese, and Butter, washed down with Tea, as opposed to Soup, Saumon Mayonnaise, Filet de Boeuf aux Champignons, Pêche Melba, Sardines en Croûte, washed down with Heidsieck (Triple sec) 1906, followed by a cup of coffee, a glass of Old Brankdy, and a "Bon Cigare". You can picture the two types of Messes, and you can almost scent the air of depression in the one, and hear the laughter in the other, with its Commanding Officer and his Officers all full of life and "bonnes histoires". Take my advice, gentlemen, and have nothing to do with the "Bully-Beef" set! Live well yourselves and enjoy your food and make all your young officers do likewise, and above all invite your General, not once, but frequently, to dinner, and see to it that you do him well. Should he be disposed to soda water rather than a glass of champagne, arrange that there is by mere chance [sic] nothing to drink except champagne! When in close touch with big towns, like Amiens, Boulogne, Doullens, Ypres, etc., see that your Mess President does not forget the fact, and when out of touch then have a standing order with Fortnum and Mason, of Piccadilly, or other equally well-known firms of "Morale-Raisers" [sic], to supply you with some of the good things that still come to these shores, in spite of the "Hun" and his "Hunnish" devices.'

Elsewhere in his lecture the General stressed that his preferred menu was a 'simple dinner that any Mess President should be able to produce', and that he had enjoyed it several times as a company and battalion commander in the Royal Irish Fusiliers during the 'hardest times in France'. He did, however, stress that his men were also getting 'all that human power could get them, e.g., their rations, plenty of beer, a good Regimental Supper Bar, and every drop of RUM [sic] obtainable, and everyone – officers and men

* This had originally been organized by Major (as he then was) J.F.C.Fuller, later the well-known military theorist and historian, in Spring, 1916, as one of the courses offered by the Third Army School in France.

– were in good heart and merry and bright in consequence, even during the hardest days of the Second Battle of YPRES [sic].'

While in reserve, battalions often found themselves having to provide working and carrying parties in order to assist those in line. In order to take some of the load off their shoulders each division was, in September, 1915, given an additional battalion, termed 'pioneers', except in the Guards Division, where it was known as the Guards Entrenching Battalion. Each division also had an Employment Company made up mainly of men who were medically unfit for front-line duty. This looked after the divisional baths, laundry and 'delousing plant', as well as the administrative elements in the headquarters and that other important contributor to morale, the 'divisional follies' or light entertainment troupe. The Company also carried out battlefield salvage.

By the end of 1915 volunteers for the New Armies had virtually dried up and the Government was forced to resort to conscription. Lord Derby's Military Service Act of January, 1916, was aimed at single men aged between eighteen and forty-one and not in reserved occupations. The demands of the war industries, however, meant that only 50,000 men registered during the first six months of the scheme, and in May universal conscription, which included married men, had to be introduced.

The conscript represented the fourth strand in the character of the British Army during the Great War. There was an understandable feeling among officers and NCOs, especially Regulars, that the conscript did not perform as well as the Kitchener volunteer or Territorial, but there is no evidence to substantiate this. True, there was a noticeable decline in morale during the second half of 1917, but this was largely brought about by war weariness and the ghastly conditions of Third Ypres, the main British offensive of that year. In contrast, the performance of the A4 Boys, of which more later, during the last six months of the war cannot be questioned. The truth was that the conscripts were assimilated into their battalions like anyone else and soon became part of the 'family'.

The Battle of the Somme, which opened on 1 July, 1916, has often been called the graveyard of Kitchener's Armies and this is true to a large extent. Only four New Army divisions took part in the Battle of Loos, but twenty-eight would be involved on the Somme. Of the seventeen divisions which took part in the opening day of the battle, there were four Regular, four Territorial and nine New Army. Looking, however, at the number of battalions that suffered over 500 casualties on 1 July, twenty out of thirty-two came from the New Armies. The 10th West Yorkshires suffered worst with no less than twenty-two officers and 688 men becoming casualties. One slightly wounded subaltern and twenty men of those who attacked came out of the battle at the end of the day. The almost total destruction of battalions in trench fighting was not new, however, and a policy had been instituted to ensure that some sixty of all ranks in each battalion were 'left out of battle' so that there was a cadre around which it could be rebuilt. Thus, the West Yorkshires would quickly regain their strength and be sent back again into the maelstrom, as the long battle dragged on. The slaughter of these battalions was harder to bear at home, especially where Pals battalions were concerned. In the towns and cities of northern England whole streets went into mourning when the casualty lists of 1 July were published, thus highlighting a disadvantage of recruiting battalions on narrow territorial lines. Against this must, however, be balanced the spirit generated when men of common background served together, and there was certainly nothing to question over the morale of the New Army battalions as they went over the top on that sunny July morning.

The Battle of the Somme lasted well on into November. It was mud and a shortage of infantry that finally brought it to a close. By October the overall British infantry strength on the Western Front had fallen from 689,000 men at the beginning of July to 576,000, and there were only 23,000 men available in the base depots in France. Every division in France and Flanders took part in the attacks at one stage or another, and the memories of the survivors are no better summed up than by Gilbert Frankau, who fought on the Somme as a field gunner, in his 1919 novel *Peter Jackson Cigar Merchant*:

'The "Somme Offensive"? What remains of it today? Only memories, bitter memories that waken men o'nights; so that they see once more the golden Virgin of Albert, poised miraculously on her red and riven tower; Carnoy shattered in its hollow, a giant baby's toy-village, dropped from careless hand and smashed in the falling; the ruins that were Mametz and the ruins that were Contalmaison and the ruins that were Fricourt and the ruins that were Pozières; see once more the crowded horse-lines blackening Happy Valley, the balloons strung out like sausages across the sky, the thousand 'planes circling like hawks above them; so that they once more hear the staccato of machine-gun fire high in the air, the dull thump of the huge and hidden naval guns at Etinehem, the roar of squat nine-point-two's on their wheelless mountings, the roar of the railway gun at Becordel, the thunder of the eight-inch and six-inch howitzers in Caterpillar Valley, the ear-splitting crunch of six-inch Mark VIIs from the road by the Craters, the manifold clamour of the Archies [anti-aircraft guns] at Montauban, the constant bark of the field guns beyond; so that they walk once more, naked and alone, among the careless ghosts of men they knew, through the horror which was Trônes Wood.'

13

Yet, in spite of the fact that the overall depth of territory gained was no more than eight miles and it cost 420,000 British casualties, it most certainly did not break the spirit of the long-suffering infantry. Furthermore, it exhausted the Germans. As Prince Rupprecht of Bavaria commented: 'What still remained of the old first-class peace-trained German infantry had been expended on the battlefield.' The coming of winter, though, gave both sides a chance to recover their strength.

The introduction of conscription in 1916 led to some rationalization of the training system. The situation had been reached whereby there were many different types of training battalion. While Special Reserve battalions continued to concentrate on maintaining the supply of reinforcements to the Regular battalions, each original Territorial battalion had formed a second battalion, initially made up of those who had not taken the Imperial Service obligation. Now that battalion had usually been given active service status and hence a third battalion had been formed to train recruits and keep the two senior battalions up to strength. Thus, the 1st/8th Londons (Post Office Rifles) had gone to France with 47th London Division in March, 1915, leaving the 2nd/8th at home. In April, 1915, however, the 3rd/8th was formed, and this took over the 2nd/8th's role at the end of that year, while the latter joined 58th Division and eventually went to France at the beginning of 1917. Kitchener, however, had in April, 1915, also established the New Armies' training on a surer footing. He converted the Fourth New Army into Service Reserve battalions and additionally ordered each locally raised battalion to form two reserve companies. These were then grouped into Local Reserve battalions. It soon became apparent, however, that counties with sparse populations were unable to supply the men to fill the battalions of their county regiments and had to find men from the more populous urban areas. It was thus decided in September, 1916, to drop the territorial titles of the New Army training battalions and merely call them Training Reserve battalions. Thus, the 31st (Reserve) Battalion Northumberland Fusiliers became the 86th Training Reserve Battalion. In July, 1917, the reserve battalions were reorganized even further, being divided into four classes reflecting the medical category, training, and age of the recruits. As a result of this the 86th Training Reserve Battalion was renumbered 273rd Training Battalion. But this was not the end of the story. Three months later the Training Reserve battalions were reincorporated into the regiments, but with no regard to their origins. For example, 258th Training Reserve Battalion (formerly 11th (Reserve) Battalion North Staffordshire Regiment) now found itself as the 51st (Graduated) Battalion, The Durham Light Infantry. All the reserve battalions supplying the Highland regiments were, at the same time, reduced to just two, 51st and 52nd (Graduated) Battalions Gordon Highlanders.

Another type of infantry battalion which was created during the war was the Garrison battalion. They were initially formed from men of medical category 'B', which made them unfit for front-line service, and were employed in guarding vulnerable parts of the British coasts and garrisoning overseas territories. A number, however, were eventually sent to France, where they were used to guard headquarters and supply dumps and to assist the Military Police on traffic control duties. One exception to this was 44th (Garrison) Battalion The Royal Fusiliers, which in 1918 manned anti-aircraft Lewis Gun posts behind the lines.

Throughout most of the war the policy was not to send soldiers to active theatres until they had reached the age of 19. To this end Provisional battalions were set up to hold trained soldiers until their nineteenth birthday. Later these were renamed Young Soldiers' battalions. Thus, Christopher Haworth joined a Training Reserve battalion on his 18th birthday in May, 1917 and, after completing his training, was then posted to Ireland. Here his battalion received orders for France and he was transferred to a Young Soldiers' battalion of the Royal Scots Fusiliers. Many of its members were youths who had been combed out of battalions in France, having enlisted under a false age, and a number had been wounded. From here he joined a home service battalion of the Argyll and Sutherland Highlanders and was, in July, 1918, eventually posted to its 14th Battalion in France.

One who did join under age was James Tait, who enlisted in the 10th Battalion East Yorkshire Regiment in April, 1915, at the age of 15. His training was first carried out in its second Depot (Reserve) Company, and he joined the Battalion *per se* on Salisbury Plain in November. Three weeks later the 10th East Yorkshires set sail for Egypt as part of 31st Division. Here the Division spent some nine weeks preparing for a Turkish invasion which never came before crossing the Mediterranean to France. After seeing service in the trenches Tait was returned to England on grounds of his extreme youth on 23 June, 1916. It would seem from his diary that he had no idea that his battalion was about to become involved in the opening day of the Somme offensive in the Serre sector and that he was lucky to leave France when he did. While delighted to be back in civilian life, he had no hesitation in enlisting again, in September, 1917, when he had finally reached the age of eighteen. He joined 7th Durham Light Infantry in France in April, 1918, as an 'A4 Boy'. These were so termed according to their medical category, which graded them as fit for active service, but as yet under nineteen years old. In the aftermath of the crisis produced by the German March, 1918, offensive the age rule was temporarily relaxed and several barely trained 18-year-olds were sent out to France to make good losses. They performed extremely well and it was on their young shoulders that the fighting through the summer and autumn largely rested. Tait himself survived the fierce fighting engendered by the third German drive in May, 1918, during which his battalion served as the pioneer battalion of 50th Northumbrian Division, and finished the war in one piece.

At the beginning of the war the minimum regulation height for recruits was 5 feet 6 inches, although by early November, 1914, this had been progessively reduced by three inches. Even so, a significant number of volunteers were turned away by the recruiting offices because they were still too short. Alfred Bigland, Member of Parliament for Birkenhead, and the main driving force behind recruiting for Kitchener's Armies in the area, was especially struck by the numbers of men who were very keen to enlist, but who were too short, and obtained War Office permission to raise a battalion of men between five feet and the minimum regulation height. The result was the 15th and 16th (Service) Battalions of The Cheshire Regiment, which formed at the end of November, 1914. Bantam battalions, as they were called, sprung up elsewhere in the country, especially in the coalfields of South Wales, industrial Glasgow, where the 18th Highland Light Infantry became known by the local people as 'The Devil Dwarfs', and northern England. There were soon sufficient to create two New Army divisions, 35th and 40th, both of whom served on the Western Front. In time, though, suitable replacements, especially in terms of chest measurement, became hard come by and 40th Division was reduced to cadre strength during the latter part of 1918.

As the war on the Western Front progressed the shape of the infantry battalion changed. In 1914 it was, apart from its two machine guns, totally reliant on the rifleman. Trench warfare, however, bred a number of new weapons and these in turn called for specialists to man them. In 1915 the Vickers Maxim machine guns had been brigaded and were taken over by the new formed Machine Gun Corps (in early 1918 the MGC companies were formed into battalions and brought under divisional control). In their place the much lighter Lewis Gun was introduced, initially on a scale of four per battalion, but later one per platoon. From early 1915 the grenade became an important weapon in the trenches; initially it was of the homemade jam tin variety, but later that year the Mills Bomb, later known as the No 36 Grenade and still in service today, was introduced. These specialities were, by 1917, reflected as low down in the organization as the infantry platoon, which now consisted of two rifle sections, a Lewis Gun section and a Bomber section. At battalion level a Lewis Gun officer and Bombing Officer advised the commanding officers and, where necessary, coordinated the activities of their respective specialists.

Another trench weapon introduced in 1915 was the trench mortar. These became categorized as heavy, medium, and light. The first two were the responsibility of the Royal Artillery, but the light trench mortar battery with each infantry brigade was manned by infantrymen, seconded from their regiments, and usually consisted of eight Stokes 3-inch mortars. They were, however, never popular with the infantry, since whenever they were deployed for a 'shoot' in the trenches, they inevitably brought down retribution from their German equivalent, the *Minenwerfer*. Sniping, too, was another speciality, with the battalion snipers and scouts controlled by the Intelligence Officer.

As has been said before, it was the offensive launched in the Ypres area on 31 July, 1917, that was the nadir in the level of morale of the infantry. Heavy reliance on artillery and unseasonably wet weather conspired to upset the delicate drainage system and the battle area soon became a quagmire. Many of the 54,000 names inscribed on the Menin Gate at Ypres as having no known grave lost their lives as a result of drowning in the mud. Siegfried Sassoon, who served with the Royal Welsh Fusiliers, was wounded twice and won the Military Cross, echoed the thoughts of many in the final couplet of his poem *Attack*:

> 'And hope, with furtive eyes and grappling fists,
> Flounders in the mud. O Jesus, make it stop'

He himself missed Third Ypres, having been sent to the shellshock hospital at Craiglockart in Scotland after writing his famous public letter in protest at the continuation of the war.

Official realization that the mind could be affected by combat only gradually dawned; in previous wars it had never been considered. Eventually special Neurasthenic Centres were set up to treat victims of shellshock and these enjoyed a significant success rate. Out of the 21,549 officers and men treated for functional diseases of the nervous system no less than 80 per cent were returned to normal duty. The main problem in recognizing genuine cases lay at battalion level, where it was up to the medical officer. He seldom had the psychological and psychiatric experience to diagnose with confidence, and was only too aware of the tightrope he walked between caring for the sick and helping to maintain the fighting spirit of the battalion. The same dilemma would face squadron commanders and their medical officers in the RAF, especially Bomber Command, during the Second World War.

Fear, though, is often infectious and it was not uncommon for officers and senior NCOs to shoot men who wavered in the midst of combat. The death sentence was also enshrined in military law and could be inflicted for a number of crimes committed 'in the face of the enemy'. Indeed, no less than 346 death sentences were carried out during the war, the vast majority in France and Flanders. Of these 266 were for desertion, three for mutiny, eighteen for cowardice, six for striking or offering violence to a superior officer, five for disobedience, two for sleeping on post, seven for quitting post, both these categories involving sentries, and two for casting away arms. The remainder were for murder.

There is no doubt that at least some of these men who were court-martialled, found guilty, and shot were genuine sufferers of shellshock. The findings of the official committee which investigated the subject after the war all but admitted that this was so and eventually, in 1930, the death sentence was dropped for crimes covering cowardice and desertion.

The Third Battle of Ypres cost the British Army 400,000 casualties, and the immediate result was Government concern that the manpower barrel would run dry. Consequently a brake was applied on the flow of replacements across the Channel. This meant that in January, 1918, there was a dramatic fall in the infantry strength in France compared to the previous January. Thus, while the administrative element on the lines of communication had grown by 400,000 men, the fighting strength had fallen by 80,000. For the infantry battalions, many of whom were by now seriously under strength, it meant that there was no way in which they could fully replace their losses. A decision was therefore taken to reduce each brigade by one battalion. The effect of this was that no less than 141 Territorial and Service battalions in France were either disbanded or amalgamated. The feelings of the 1st/8th Londons, who were told to disband, their A and B Companies being sent to the 1st/17th Londons and C Company to the 1st/24th Londons in the same division, while the remainder were absorbed by the 2nd/8th in 58th Division, are described by Ernest de Caux, their interpreter, who, with the Commanding officer and Quartermaster, represented the only Battalion officers who had been with it when it first arrived in France in March, 1918:

'A farewell dinner took place on 29th January at Battalion Head-quarters, one of the simplest meals ever served in that hospitable mess: soup, beef, peas and mashed potatoes, tinned peaches. Only the older members of the Battalion had glasses, the remainder cups and odd crockery. Boxes and benches served as seats. While the port was being passed round the Last Post sounded and during the hush that fell upon the company while the solemn notes were ringing out Colonel Vince rose. He recalled the Battalion as it had arrived in France in the pride of its strength; and its losses. Now there would be no collective home-going. . . . All that remained to the survivors was the memory of their comrades. . . . While he was speaking the hum of Gothas [German bombers] sounded overhead and the warning whistle of the guard. The rattle of machine-guns, the bark of anti-aircraft and the crump of bombs formed a crescendo to the measured heartfelt words of what in the truest sense was the funeral oration of the Post Office Rifles. As Colonel Vince sat down the brazen notes of a bugle somewhere in the village sounded the Last Post once more. In the silence that ensued sounds of laughter and song from the Nissen hut across the yard broke the spell and thither Battalion atmosphere. The incandescent character of the coke brazier and the libations alike had raised the temperature to white heat. What cared the inmates for the second bombing that like a thunder-storm gathered, broke and passed, while listening to the cobbler-sergeant's stories; or the reminiscences of Company Quartermaster Sergeant Thompson, a smooth-faced little man who had managed to survive since the days of Festubert [May, 1915] with the same cheerful smile. There was [Regimental] Quarter-master Sergeant Brand MM, who had dragged bags of rations across the craters of Vimy Ridge in daylight and Sergeant-Major Johnstone, of Messines Ridge and Bourlon Wood. And the craftsmen who had done odd jobs in very queer places. Their history and the history of those who had gone before them was the true history of the Battalion.'

This does more than hint at what Regimental Spirit is all about. It was those in the battalion who made its history and however heavy its casualties there would always be some survivors to hand on the torch, and to also act as the repository for what had gone on before.

Those who found themselves posted to other regiments and battalions quickly assimilated into their new families. This was helped by the fact that the indicators were growing that the Germans were planning an offensive. When it broke on the fog-laden early morning of 21 March, 1918, it was to prove a severe test of character for the infantrymen of the Third and Fifth Armies, especially the latter. Many fought on in their strongpoints, bypassed and then surrounded, until their ammunition had been expended or further resistance was useless. The remainder fell back, but after a week's desperate fighting the situation began to stabilize.

The German attacks finally ceased on 5 April, but four days later they struck again, this time south of Ypres. So serious was the situation that the normally dour Haig issued his famous and emotive 'Backs to the Wall' message, but once again the defence held, although further ground was lost. There was then a short respite before the Germans struck again. Five British divisions, exhausted by the fighting in March and April, had been sent there to recover since it was considered to be a quiet part of the front. Now they found themselves engulfed once more. During the battle the 2nd Devons particularly distinguished themselves, and, indeed, their performance typifies that of many battalions during the critical period of spring 1918. They were ordered to hold a feature called the Bois du Buttes in order to cover the withdrawal of the remainder of 8th Division. Subjected to a heavy bombardment and then engulfed by German storm troops, they held on until the last, but bought vital time for defences south of the River Aisne to be prepared. Lieutenant-Colonel Anderson-Morshead, whose coolness had done so much to inspire his men, was killed, but the battalion fought on, even with their bayonets after their ammunition had run out. Twenty-eight officers and 552 other ranks were killed, wounded or captured. In the words of General Berthelot, commanding the French Fifth Army, under whose command the Devons were, they 'willingly made the supreme sacrifice demanded by the sacred cause of the Allies'. Further French acclaim came in December, 1918, when the reconstituted battalion was presented with the Croix de Guerre, still worn with pride by today's Devon and Dorsets.

Two more German offensives took place, in June and July, again against the French, before their bolt was finally shot. Now it was the Allied turn, but, in spite of the exhortation of Generalissimo Ferdinand Foch of '*Toute La Monde á la Bataille!*' few believed that the war would end before the end of the year. Yet it was so, but not without hard fighting as the Germans were slowly pushed back. Beginning with the attack at Amiens on 8 August, which was spearheaded by tanks, Australians and New Zealanders and which Ludendorff called 'the Black Day for the German Army', the BEF took a leading part, advancing seventy miles on a hundred-mile front and capturing more prisoners and guns than either its American or French allies. Casualties, though, were heavy, with 350,000 killed, wounded or missing. One young officer, who was wounded on 1 September and rejoined his battalion on the day before the Armistice, noted that it was 'like joining a new battalion'. It also took the infantry time to get used to open warfare. The specialist needs of the trenches meant that standards of musketry had slipped alarmingly and drills for coordinating the various arms in quick attacks had to evolve virtually from scratch. As one divisional report succinctly put it: 'Owing to the lack of training fire control and fire discipline was [sic] absent and fire to support movement rare in application.'

When the guns did finally cease fire at 11am on the grey morning of 11 November, it came as a surprise to most. The 5th Highland Light Infantry:

'At 07.00 on the 11th November we set out for our last attack, our objective being the Mons-Jurbise road. There was no opposition of any kind and by 09.00 we had reached the objective. Our job had proved an easy one, and we quite expected to get orders to continue the pursuit. But of a sudden there arose a clatter of hoofs and an obviously excited transport officer dashed up to the Commanding Officer, brandishing one of those pink forms we had learned to hate. But never before had an Army Form borne such a message as this:

"Hostilities will cease at 11.00; until further orders units will not move beyond the position occupied at that time." At last there had dawned the day for which we had lived – and so many had died. Strange to relate there was no tremendous excitement.'

James Jack had crossed to France as a platoon commander in the 1st Cameronians in August, 1914, commanded a company and been second-in-command of the 2nd Scottish Rifles during 1915–16, taking over the 2nd West Yorkshires in September, 1916, and commanding them with distinction for almost a year before being severely wounded on the opening day of Third Ypres. Recovering from his wound, he then briefly commanded the 1st Cameronians before being promoted to brigade commander for the final Push. For him Armistice Night was passed being entertained to dinner by the mayor of the small Belgian town in which his headquarters found itself at the end of hostilities:

'At last I lie down tired and very happy, but sleep is elusive. How far away is that 22 August 1914, when I heard with a shudder, as a platoon commander at Valenciennes, that real live German troops, armed to the teeth, were close at hand – one has hardened since then. Incidents flash through the memory: the battles of the first four months: the awful winters in waterlogged trenches, cold and miserable: the terrible trench-war assaults and shell fire of the next three years: loss of friends, exhaustion and wounds: the stupendous victories of the last few months: our enemies beaten to their knees.

Thank God! the end of a frightful four years, thirty-four months of them at the front with the infantry, whose company officers, rank and file, together with other front-line units, have suffered bravely, patiently and unselfishly, hardships and perils beyond even the imagination of those, including soldiers, who have not shared them.'

No finer tribute could be paid to the British infantry on the Western Front 1914–18.

CHAPTER TWO

The Other Theatres
1914–1918

WHILE the Western Front occupied the centre stage position for the British Army through 1914–18, there were many other theatres in which infantry battalions found themselves involved in varying capacities. One of the most exotic and remote campaigns occurred during the opening months of the war and earned the 2nd South Wales Borderers the unique battle honour of 'Tsingtao'. This was the port of the German-leased territory of Kiaochow on the China coast, and the Japanese had long had their eye on it. Within two weeks of the outbreak of war Japan, after a request by a British government fearful of the German threat to Far Eastern maritime trade, declared war on Germany, and before August was out had dispatched a force to seize Kiaochow. The South Wales Borderers were part of the garrison of the treaty port of Tientsin and were sent to join the Japanese, together with the 36th Sikhs, at the beginning of September in order to display Allied solidarity to the Germans. Tsingtao eventually surrendered on 7 November, but for the British contingent it was a frustrating experience. The Japanese were unwilling to give them more than a very subsidiary role and transport difficulties meant that they were ill-equipped for siege warfare. The weather was appalling, with persistent heavy rain, the troops having only thin khaki drill uniform and being supplied with no tentage. Rations, too, were woefully inadequate. Given these conditions, the sick rate remained remarkably low.

The 2nd Loyals had an even more depressing experience. They found themselves as part of a hastily organized force of low grade Indian troops which was sent across the Indian Ocean to established a toehold in German East Africa at Tanga, just north of Zanzibar. A month at sea, in very cramped transports, left the troops debilitated and, although they managed an unopposed landing outside the town on 2 November, they found themselves totally ill-equipped to cope with an advance inland through the bush to capture Tanga itself, especially when faced by the highly skilled German-led native Askaris. Angry swarms of bees, their nests disturbed by small arms fire, did not help. The result was an ignominious withdrawal and re-embarkation.

The Loyals returned from Mombasa to the fray in Spring, 1915. This time they were in the company of the 25th Royal Fusiliers (Legion of Frontiersmen), one of the more unusual New Army battalions. The Legion itself, which still exists today, had been formed in January, 1905, by Lord Lonsdale, the 'sporting Earl', and Roger Pocock, an ex-Canadian 'Mountie' turned travel writer and journalist, as a volunteer body to help defend the frontiers of the British Empire, attracting many who had 'roughed it' in remote corners of the world. The 25th Royal Fusiliers itself had been formed by Colonel Daniel Driscoll, who had raised and commanded a body of scouts during the Boer War, and was made up Frontiersmen types. Prominent among them was Captain F.C. Selous, famed white hunter and explorer, who was killed in action in January, 1917, at the age of 65. His name, however, lived on in the Selous Scouts, who were white Rhodesia's spearhead in its eventually fruitless counter-insurgency campaign against the black liberation forces in the 1970s. Others included American cowboys and an American millionaire who owned a large farm in East Africa. There were also some Russians who had escaped Siberian exile, veterans of the French Foreign Legion, and even a former Honduran general. Some, though, seemed ill suited, at least on the surface, for this band of desperadoes. A noted photographer, an opera singer, a circus clown, a lion tamer, who admitted being afraid of lions, a lighthouse keeper from Scotland, and a Buckingham Palace footman were among the members of the Battalion.

These two battalions were now to find themselves involved, alongside South African, Indian, Rhodesian troops, and the King's African Rifles in a long and frustrating campaign during which they never would succeed in running the wily German commander, Paul von Lettow-Vorbeck, to earth. The often rough terrain and a shaky supply system did not help, and there was also the problem of disease. Malaria, blackwater fever, dysentery and chiggars, caused by fleas burrowing in human flesh, were omnipresent. The Loyals had a strength of 900 men in March, 1915. Three months later this had been reduced to under a third, almost entirely due to sickness. Although brought back up to strength by fresh drafts, the Battalion had to be sent to South Africa to recuperate for three months in April, 1916, but by the end of the year could again only muster 345 fit men. Indeed, the proportion of sickness to battle casualties was a staggering 30:1. As one 25th Royal Fusilier wrote: 'Ah, I wish to hell I was in France! There one lives like a gentleman and dies like a man. Here one lives like a pig and dies like a dog.'

The Middle East, too, quickly became an active theatre of war. Here

Turkey presented two initial threats. The first was to the British oilfields around Abadan at the head of the Persian Gulf. Two Indian infantry brigades were successively landed here in early November, 1914. They included two British battalions, the 2nd Dorsets and 2nd Norfolks, and early successes against small garrisons led to the precipitate Turkish evacuation of Basra, which the force entered on 21 November. Encouraged by this, the India Office, which was, overall, responsible for the region, now believed that Baghdad was ripe for quick seizure. Accordingly, D Force, as it was called, advanced up the right bank of the Euphrates, seizing the important strongpoint of Quirna, which stands 45 miles north of Basra and at the junction with the Tigris, on 8 December. The seasonal floods prevented any further movement forward, and when Spring came it was the Turks who moved first, against the main British base at Shaiba, which lay west of Basra, but had been cut off from it by the floods. The British attacked and routed them in what was probably one of the last battles in which British infantry officers drew their swords in anger. The advance on Baghdad itself began at the end of May, assisted by a flotilla of gunboats steaming up the Tigris as the troops advanced along the bank. In temperatures which averaged 113°F at midday and with water always short, even given the proximity of the river, Force D pushed on towards its next major objective, Al Amarah, which lay 60 miles north of Quirna. This was captured on 3 June and it seemed that Turkish resistance was everywhere collapsing. General Nixon, commanding D Force, was all the more determined to continue to Baghdad. The Viceroy of India was keen, but Whitehall had its doubts.

Nevertheless it was agreed that Nixon could continue north to the next intermediate objective, Kut, also on the Tigris, but 90 miles to the north-west. During the pause after the capture of Al Amarah, General Town-shend, commanding the advance itself, went back to India on three months' sick leave and the Turks prepared comprehensive defences at Kut. The advance itself resumed on 12 September, still accompanied by the naval flotilla, and with temperatures continuing high. Two weeks later Town-shend attacked and seized Kut, although command and control problems prevented him from cutting off the Turkish defenders, who withdrew northwards. The debate over whether to continue to Baghdad then raised its ugly head once more, but again Nixon got his way, even though Townshend, his lines of communication ever lengthening, had expressed his concern, especially as he had learnt that the Turks were reinforcing. At the end of October Townshend pressed on again and came on the Turks at Ctesiphon on 21 November. The following morning he attacked and managed to gain the Turkish first line of trenches, but could go no further in the face of determined counter-attacks. The result was stalemate, and with reports coming in that the Turks were bringing up more men, Townshend decided that he must withdraw. One British soldier's wry comment on the battle was: 'Of the officers, some calls it Tesiphon and some calls it Sestiphon, but we calls it Pistupon.'

Followed up by the Turks, Townshend fell back to Kut al Amara. Here Nixon ordered him to hold and by 7 December the Turks had the town surrounded. The garrison of 12,000 men, including 2,000 sick and wounded, began the siege with morale high. Nixon stated that they would be relieved within two months, and stocks seemed more than sufficient. Besides, Charles Townshend himself was none other than the commander of Chitral during its epic siege in 1895 (see Vol I). The British infantry was represented by the now Mesopotamian veteran 2nd Norfolks and 2nd Dorsets, together with the 1st Oxfordshire and Buckinghamshire Light Infantry, half of the 2nd Royal West Kents, and a company of the 1st/4th Hampshires, the only non-Regular element. By now the Indian Corps had arrived in-theatre from France, and was to form the Kut relief force. Setting out from Amara on 3 January, the relief force advanced on both sides of the Tigris, coming up against the first blocking position on the 6th. Resistance was still stiff, and at the end of two days' fighting the Turks continued to hold. Worse, casualties were heavy. The 1st Seaforths lost twenty officers and 360 men, and the 2nd Black Watch, also in the 7th (Meerut) Division, suffered likewise. The Turks withdrew the following day, but the 4,000 casualties that they had inflicted could be ill spared. Worse, it took eleven days before the wounded could be evacuated from the battlefield; they were left, most of them with their wounds still bound with their field dressings, lying in the mud. The medical system had completely broken down and many of these unfortunates died needlessly from dysentery.

On 13 January the relief force came up against another position, and a further one a week later. Casualties were again high, and matters were made worse by the fact that the British were unable to force the third position. Again the wounded suffered, and the Black Watch were reduced to a mere two officers and fifteen men. These were absorbed by the Seaforths, and the combined unit became known as The Highland Battalion.

By now Nixon had been replaced by Lake, who decided that the relief force must continue its efforts, rather than Townshend attempt a break-out. The fourth attempt at a decisive breakthrough came at the beginning of March, but once more ended in failure. Further efforts in April met with the same result. During the period January–April, 1916, the relief force, otherwise known as the Tigris Corps, suffered 23,000 battle casualties, with the Highland Battalion losing 921 men during the period 4–22 April alone.

Meanwhile, in Kut itself the initial optimism quickly degenerated after the early failures of the Tigris Corps. Rain, frost and lack of fresh vegetables led to an ever sharper rise in the sick list. Then, in March, food began to run low, until by early April the garrison was subsisting on a daily ration of five ounces of grain and a little horseflesh. Soldiers began to faint from hunger. A few supplies were delivered by air, but they merely served to prolong the agony. Faced now with total starvation, Townshend saw that there was no realistic option but to open negotiations for surrender. On 29 April the garrison marched into captivity. While the 1,450 sick and wounded were quickly exchanged, the 12,000 remaining POWs were to suffer much hardship during the next two and a half years. No less than 4,000 would die, including 70 per cent of the British rank and file. Indeed, only thirty 2nd Dorsets survived. Eventually the disgrace of Kut would be avenged and Baghdad would fall, in March, 1917, and the British would finally enter Mosul in November, 1918.

The campaign in Mesopotamia continued to be regarded somewhat of a backwater, and the climate and terrain, as well as long lines of communication, meant that it remained a tough one. One footnote to the campaign was the expedition led in early 1918 by Major General L.C. Dunsterville, the Stalky in Rudyard Kipling's *Stalky and Co*, from Baghdad to help prevent the Turks from taking advantage of the vacuum created by the Russian Revolution and seizing Baku and its surrounding oilfields on the Caspian Sea. His small force was built around the 7th North Staffords, and included detachments of the 1st/4th Hampshires, 9th Worcesters, and 9th Royal Warwicks. They were eventually forced to destroy the oilfields and found themselves under siege. They were now ordered by London to withdraw, which they did under fire of the Bolshevik fleet. The Russian crew of one of their transports were, however, so impressed with the performance of Dunsterville's men that they presented him with a petition praising their 'heroic conduct' and asking that they themselves be granted British nationality. The North Staffords, however, were to return to Baku in November after Turkey and Germany had signed armistices with the Allies, albeit just for a short time.

The second initial Turkish threat was to the Suez Canal, and it was for this reason that the 42nd (East Lancashire) Division (TF) was sent to Egypt in September, 1914. Here it joined Indian Army formations and in early December the forces in Egypt were reinforced by troops from Australia and New Zealand. The Lancastrians were bemused by their colonial cousins, not just because of their more relaxed discipline, but also the reputation they quickly achieved of being wild men, who did their best to tear Cairo apart during the Christmas season. The Australians themselves blamed this on a small minority of old soldiers who had fought in the South African War or served in the British Army, and sent the worst troublemakers back to Australia. The Turks did make some half-hearted efforts to cross the Canal in early February, 1915, but were repulsed with little difficulty by the Indians, who were manning the forward positions while the Lancashire Territorials and Anzacs remained in reserve. These troops would, however, soon face a much sterner test.

The concept of seeking a decisive victory over the Central Powers in a theatre other than France and Flanders began to evolve before the end of 1914, once the fighting at Ypres had died down and it had become clear that no early Allied victory would be possible in the West. Winston Churchill, then First Lord of the Admiralty, was one of the prime movers and early in the new year a request from Russia for active support to relieve Turkish pressure on her in the Caucasus made the idea seem even more attractive. Consequently, plans were set in train for a joint Anglo-French expedition to force the Dardanelles Straits, which connect the Black Sea with the Mediterranean, land a force, and press on to Constantinople, thereby, it was hoped, knocking Turkey out of the war. Thereafter the possiblities of advancing up the Balkans and then, in conjunction with the Russians, striking Germany and Austro-Hungary from the east appeared to promise an early victorious ending of the ear. Initially the one major obstacle that stood in the way was Kitchener, who complained that he did not have sufficient trained troops to spare. Consequently the Royal Navy produced a plan for the methodical destruction of the Dardanelles forts, with the fleet then threatening Constantinople with bombardment. Kitchener now relented, but not before the Anglo-French fleet had bombarded the forts and attempted, without much success, to clear the minefields in the Narrows. A further attempt was made the following month, March, but resulted in the loss of three battleships sunk and three crippled. All this merely served to alert the Turks.

Kitchener, having relented, gathered a force, which consisted of the 29th Division, formed from Regular battalions which had been brought home from India, the Australians and New Zealanders in Egypt, who were formed into the Anzac Corps, and the Royal Naval Division, formed at the outbreak of war from sailors surplus to the fleet, and which had already taken part in an ill-fated attempt to reinforce Antwerp before the Germans took it in Autumn, 1914. French troops would also take part.

The force assembled at the island of Lemnos, which became the last resting place of Rupert Brooke, the poet who so accurately reflected the mood of Britain in the early months of the war. Many who had received a classical education saw themselves as latter-day Greeks sailing to take part in the Trojan War and there was a belief that their efforts would prove decisive in ending the war. Major-General Aylmer Hunter-Weston, who had already fought in France, took a more sanguine view. In a personal message to every member of the 29th Division he warned them to 'be prepared to suffer hardships, privations, thirst, and heavy losses, by bullets, by shells, by mines, by drowning'.

The plan was for the force to land on the Gallipoli peninsula on the Aegean side of the Narrows, with the Anzacs assaulting the west coast, while 29th Division landed on the tip of the peninsula. The landings took place on 25 April. The Anzacs landed north of their intended beach, but initially took the Turks by surprise. As they advanced inland, however, the broken ground and their inexperience resulted in growing confusion, as units became hopelessly jumbled and this gave the Turks time to recover from their initial surprise and deploy reserves. The 29th Division was allocated five beaches, S to Y, and there was to be a one-hour naval bombardment before they went ashore. W and V Beaches were on the very tip of the peninsula, and the 1st Lancashire Fusiliers were to land on the former. The Turks had suffered heavy casualties during the bombardment, but they continued to man their trenches and were under strict orders not to open fire until the attackers actually landed. They also noted, with satisfaction, that the naval gunfire had left the barbed wire on the beaches intact. Thus, as the ships' boats carrying the Lancashire Fusiliers ground to a halt on the foreshore and the first men jumped out, a heavy fire was opened. A midshipman, commanding one of the boats, commented that it was 'a most extraordinary thing how many of the cutters which were open and absolutely unprotected and crambed [sic] with men survived it. The soldiers had to jump into about 3 feet of water and it was a most awful sight seeing them being shot down as soon as they got into the water and the wounded ones hanging onto the edge of the boats. Lots of slightly wounded men must have been drowned.' Those who did get ashore were now faced by the wire and many more fell as they desperately tried to cut their way through it. A few small groups eventually managed to get through it and sought cover in the sand dunes beyond. The slaughter had, however, been immense, with no less than 533 Lancashire Fusiliers killed and wounded, but their exemplary courage in the face of so much adversity was to be rewarded by six Victoria Crosses.

Matters on V Beach were even worse. Here 2,000 men of the 1st Royal Munster Fusiliers, 2nd Hampshires and 1st Royal Dublin Fusiliers steamed into the shore on a hastily converted steamer, the *River Clyde*, the idea being that she would run aground and a bridge of boats would then enable the attackers to reach the beach dryshod. One officer on board noted in his diary: '0622 hours. Ran smoothly ashore, no opposition. We shall land unopposed.' But here again the Turks, a mere three platoons' worth with four machine guns, held their fire, only opening it as the attackers began to run down the gangway of the *River Clyde* to the waiting boats. The result was another slaughter, with only 200 men out of 1,500 who attempted to land getting ashore. In contrast, the 2nd Royal Fusiliers made an unopposed landing on Y Beach and met only slight opposition, which they easily overcame, on X Beach. The same happened to the 2nd South Wales Borderers, who were seeing their first action since Tsingtao, on S Beach. Sadly, what success had been achieved was not exploited, largely because the senior commanders were still afloat and ship-to-shore communications were poor. There was also a marked reluctance to adjust previously drawn up plans to reflect the real situation as it now existed.

Thus the campaign in the Dardanelles now degenerated into trench warfare, in conditions that were even tougher than on the Western Front. The front lines were generally very much closer to one another and there was nowhere on the Peninsula which was out of range of shellfire. Attacks across the trenches often withered in their tracks before they had got properly underway. A subaltern in the Royal Warwicks attached to the 1st Royal Inniskilling Fusiliers describes what happened to him during the Third Battle of Krithia in early June:

'4-6-15 Big advance ordered. Strong points bombarded between 8–10.30am. 10am moved into support trenches under heavy shellfire. Lieut Spragg killed by shrapnel. Hit in the head & heart. Moved along support trenches during our heavy artillery bombardment. Platoon Sergeant Carlin and many others wounded. Arrived in first line trenches. Ordered by Capt Gilbert to advance over the parapet with my platoon. This I did under a very heavy rifle and machine-gun fire and lay down about ten yards out. One man hit on my right through the chin getting down. He managed to crawl back to the trench. I found the man on my left was dead. After waiting here for about twenty minutes unable to get in touch with anyone on either flank and having only five men left I decided to crawl back to the trench and jump in. This we did.'

The heat of the summer brought flies, and they, the open latrines, unburied bodies of both men and animals, impure water, and the monotonous diet combined to make dysentery widespread. Lice were another problem, although this was common to the Western Front as well. One NCO wrote: 'We itched and scratched until we were tired with scratching; we turned our clothes inside out and ran the burning ends of cigarettes up the seams. The crackle of a frizzled louse was one of the sweetest sounds we knew.'

As the summer wore on more and more reinforcements poured into the Peninsula. The Lancashire Territorials from Egypt were among the earliest, and they were followed by three New Army divisions – 10th (Irish), 11th (Northern), 13th (Western) – and three further Territorial Divisions – 52nd (Lowland), 53rd (Welsh), 54th (East Anglian). Indian troops also fought here. These reinforcements were, however, in vain, for deadlock now existed, and those in the Dardanelles became increasingly conscious that they were now in a backwater. Worse, on 26 November there was a severe thunderstorm, with gales and torrential rain. This flooded the trenches and was immediately followed by a blizzard. Caught totally unprepared, the troops suffered 280 deaths from exposure and no less than 16,000 frostbite casualties. It was the last nail in the Dardanelles coffin and the decision was taken to evacuate. This was carried out during the last days of December and early January. Elaborate deception measures resulted in what was the most successful operation of the whole campaign, with not a man being lost. Few were sorry to leave.

The Dardanelles divisions now returned to Egypt. The Anzacs and 29th Division were sent from there to France, in time to take part in the Battle of the Somme, where the 29th was to suffer heavily on the first day in the attack against Beaumont Hamel. The 10th (Irish) Division, on the other hand, found itself in Salonika. The decision to send Anglo-French troops there had been taken in Autumn, 1915, as a means of giving direct support to hard-pressed Serbia and as a substitute for the failure in the Dardanelles. With Serbians, a Russian brigade and Italian troops also becoming involved, it became the most cosmopolitan of the Allied theatres of war. The enemy were the Bulgarians, stiffened by Germans, but the mountain-ous nature of the terrain meant that little could be achieved until the very end of the war, even with the steady growth of the size of the Allied force, with a further five British divisions (22nd, 26th–28th, 60th) seeing service there. Not for nothing did the Germans gleefully call it the 'largest concentration camp'.

The remaining Dardanelles veterans were committed to the campaign to drive the Turks out of Palestine. The early attempts to break through the Turkish lines at Gaza failed, and it was only after Allenby relieved Murray in command of the Egyptian Expeditionary Force in July, 1917, that progress was made. Allenby received reinforcements in the shape of 10th (Irish) and 60th (2nd/2nd London) Divisions from Salonika,and two fresh divisions from Britain, the recently raised 74th (Yeomanry) and 75th Divisions. The 74th, as its subtitle suggests, was formed from surplus yeomanry regiments, who were rebadged as battalions of the Line, but wore a broken spur as their divisional emblem as a reminder of their antecedents. The Third Battle of Gaza opened on 31 October with the capture of Bathsheba by double envelopment. There then came the break-in of the main Turkish defences around Gaza. It was the cavalry of the

Desert Mounted Corps who were supposed to enjoy the glamour of the victorious pursuit, but shortage of water severely restricted them and it was the infantry who had to take on the main role. The Turks, though, put up stubborn resitance during the advance to the line Jaffa-Jerusalem and there were a number of tough actions. The experience of the 5th Highland Light Infantry is typical. They set out twenty-nine officers and 699 men strong, but by 1 December had lost twenty-five officers and 368 men, killed, wounded and sick.

'In the three weeks the men had not ceased from fighting and marching. They had often been on half rations, without tobacco or home mail, never sufficiently clad and without any real rest or sleep. The fighting had been mainly night attacks, over unknown and unseen ground, to positions that had to be located by the enemy's fire. Nothing tries troops like fighting in pitch blackness. Every man is a law unto himself and the only things that tell are grit and discipline. The Battalion might be weary and footsore, hungry and tired, battle-weary and nerve-wracked, yet the men always had that little reserve of heart left which lifted them through the most trying day or the deadly night.'

The climax came with the surrender of Jerusalem. The Turks eventually evacuated it and the mayor was desperate to surrender the city before it suffered damage.

It was the Londoners of 60th Division who had Jerusalem as their ultimate objective, and the mayor, bearing a white flag and the keys of the city, first attempted to surrender to some cooks of the 2nd/20th Londons who were lost. They declined, probably with much embarrassment, as did two sergeants commanding an outpost of the 2nd/19th Londons, and even some Gunner officers, who were much more concerned in getting their guns in action against the Turkish rearguards, although they did pass a message back by telephone. Eventually it was the commander of 60th Division, General Shea, who accepted the surrender on Allenby's behalf, and honour was further satisfied when Allenby himself made his formal entry into the Holy City on foot.

After this there was a lull, especially with the winter rains, which slowed down movement, and the need to run the railway and water pipeline forward in order to support future operations. These were resumed in February with an advance up the Jordan valley, the rugged country and the rains, which unusually lasted well into March, making life very difficult. Then came the German March, 1918, offensive in France. Urgent demands for reinforcements meant that Allenby had to send the 52nd and

74th Divisions complete to France, together with an additional twenty-four infantry battalions. For those involved, the thought that they were moving 2,000 miles closer to home, and might have the chance of leave and seeing their loved ones for, in the case of some, the first time in three years, was balanced by the Western Front's reputation as a 'mincing machine'. In their place Allenby received two Indian divisions from Mesopotamia and additional battalions from India. This meant that he had to totally reorganize his four remaining British infantry divisions, the result being that each was left with just one British battalion per brigade. This reorganization took time and it was not until September, 1918, that the Egyptian Expeditionary Force was ready for what would be the knock-out blow. This came at Megiddo, where the infantry once more achieved the break-in and the cavalry finally had the opportunity of galloping through the 'Gee in Gap' about which they had dreamed for the past four years.

Still closer to home the infantry found themselves involved in another theatre in Autumn, 1917. This was as a result of the Austro-German offensive at Caporetto, which drove the Italian armies back some 85 miles to the River Piave. A British and a French corps were sent from France to bolster their now demoralized ally. The British contingent consisted of the 5th, 7th, 23rd, 41st, and 48th Divisions, although the 41st returned to the Western Front in Spring, 1918. Their first main task was to help repulse the final Austrian offensive, aimed at forcing a passage across the Piave, in June. Then, in October, came the Allied assault across the Piave, what became known as the Battle of Vittoria Veneto and which provided sweet revenge for the Italians for their reverse of twelve months before. As one of the preliminary operations, a brigade of 7th Division had to capture an island in the middle of the river, which was swollen by heavy rain. All that were available to make the crossing were twelve Italian scows, each of which could carry only seven infantrymen besides its two boatmen. The skill of the latter and the determination of their passengers enabled all objectives to be secured by dawn. The defence broke and the 48th Division led the advance from the Asiago Plateau into the Austrian Tirol and claimed to be the only Allied troops fighting in Western Europe to enter enemy territory before the armistice was signed. Their crowning moment was accepting the surrender of a complete Austrian corps.

Finally, there were the three Territorial divisions (43rd–45th) which were sent to India in Autumn, 1914, in order to relieve Regular British battalions stationed there, and other Territorial Force battalions which were sent to such places as Aden, Cyprus, Gibraltar, Hong Kong and Malta to fulfill the same purpose. Many of the Territorials involved soon began to feel that they had been penalized by volunteering early for foreign service and then being shunted into these backwaters. Some battalions did, it is true, eventually get sent to France and Mesopotamia, but others remained in India throughout the war. Worse, a number of Territorials about to return to Britain in 1919 found their passages home cancelled and became caught up in the Third Afghan War, of which more in the next chapter. Even so, the Territorials in India did serve on the Frontier during the war, including during the Mohmand uprising of 1915 for which they became entitled to the Indian General Service Medal.

Thus, the Great War found the British infantryman engaged in many peripheral campaigns of one sort or another, often in inhospitable terrain and in extremes of climate. Once more his sense of duty, courage and endurance were challenged, but he met it with the same stoicism that his predecessors had done for over 250 years. The coming of peace, after a war in which he had suffered probably as in no other, would not, however, mean an end to the challenges that were laid before him.

CHAPTER THREE

Back to Empire
1919–1939

AT the end of 'the war to end all war' the British Army had a staggering overall strength of just under 3,800,000 men. The vast majority had enlisted or been conscripted just for the duration of the war and uppermost in their minds was demobilization. This, though, had to be balanced against the vastly increased commitments that the Army now faced, at least until the details of the treaties which would formally bring the war to an end could be worked out. Troops had to be sent to occupy the Rhineland and, indeed, began their march eastwards within a few days of the Armistice. The conquered Turkish territories in the Middle East also required large garrisons, and there was still the Empire to defend.

There were two additional complications. Britain and her allies had found themselves embroiled in the Russian civil war, initially in an attempt to keep German troops tied down in the East, but, after November, 1918, primarily to support the Whites in their attempts to overthrow the Revolution. Much closer to home there was also a deteriorating situation in Ireland.

In the light of all this it was clear to the Government that they could not immediately reduce the Army to its prewar size, but on 11 November, 1918, military conscription had at once been suspended. At the same time demobilization had to go ahead. An elaborate plan had been drawn up, its main object being the avoidance of mass unemployment. To this end priority was given to men in key trades and industries, whom it was hoped would quickly develop jobs for the others. Employers were allowed to select who should be given early demobilization and tended to opt for those who had been conscripted last. All this quickly resulted in growing discontent among the troops, especially those who had enlisted early in the war. There was demonstrations, verging on mutiny, by men in transit camps on the South Coast of England who were returning from leave to France and fears grew that Bolshevism might be taking hold. Prime Minister David Lloyd George therefore appointed Winston Churchill Secretary of State for War in January, 1919, and he immediately scrapped the scheme and gave priority to those who had enlisted before 1916. He also brought back the Guards Division from France. It paraded through the streets of London as a signal that the Government would no longer tolerate unrest by disaffected soldiers and others. As for trying to maintain the Army at sufficient strength to meet its commitments, the War Office introduced a scheme in December, 1918, whereby men could re-enlist from two to four years with bounties of £20–£40. The following month the minimum age of enlistment was lowered from eighteen to seventeen and Churchill also doubled the rates of pay. Finally the Naval, Military and Air Force Service Act, passed in March 1919, extended conscription until the end of April, 1920.

The New Army battalions garrisoning the Rhineland were quickly disbanded, while the Territorial battalions returned home. Their place was taken by Young Soldiers and Regular battalions. After the Treaty of Versailles had been signed the garrison was reduced and eventually withdrawn in 1929, little knowing that sixteen years later there would be a much larger garrison in place and that it would stay in Germany for a considerably longer time.

Turning to more active regions of the world, the first infantrymen to become embroiled in the Russian Civil War did so some months before the Great War came to an end. First to become involved was the 25th Middlesex Regiment, a garrison battalion which had served in India prior to being sent to Hong Kong. Here its lotus life was suddenly interrupted in July, 1918, when it was ordered to take ship to Vladivostok. Here the Battalion was supposed to join American, French and Japanese contingents in supporting the Czech Legion, which, having fought its way eastwards along the Trans-Siberian railway was now seen as the principal anti-Bolshevik force in Siberia. The Middlesex, all of low medical category and lacking tents and mosquito nets, duly landed and quickly found themselves in the front line. Here they were placed under Japanese command, and harboured the same suspicions that the South Wales Borderers had at Tsingtao in 1914, namely that the Japanese wanted any glory for themselves. Differing aims among the Allies, however, meant that, having been involved in one successful brush against the Reds in August, the Middlesex were restricted to garrison duties and their standing, especially with the White Russians and Czechs, fell. They were therefore more than happy to be relieved before the year was out by the 1st/9th Hampshires. This battalion had come from India via Colombo and Hong Kong, with several men falling victim to the global influenza epidemic while en route. On arrival at Vladivostok they were ordered by train to Omsk, which took them by way of Manchuria in temperatures as low as −40°. They appear to have adapted very quickly. One of their NCOs, Sergeant Tulley, wrote of their journey: 'On the whole we are having a real good time of it, and the food is quite good. For instance, our breakfast this morning was bacon, porridge, coffee, bread, butter and jam, while for Christmas dinner we had geese, bacon, peas, potatoes and a duff pudding.' Arriving at Omsk in early January, they immediately impressed the local population. A local newspaper reported their arrival:

'They march – if march is the word – with the light feet of sportsmen. Excellently clothed, healthy, fresh. The faces of strong determined people, expressing the blood of the whole nation. All seem young and at the height of their strength. . . . Everything is strange to them, yet they walk about the town as if they had lived here all their lives, quietly, without open hostility, nor do they lose themselves as foreigners do. Their officers are simple and quiet, happily good-hearted, noble. . . . Yes, a fine people – what a fine people the English are! These Hampshires make us feel that about them.

Here in a nutshell is the British soldier's ability to adapt quickly to a new environment. The Hampshires, like the Middlesex, would, however, find themselves as mere bystanders, and were withdrawn at the beginning of November, 1919, sailing home via Vancouver.

In other parts of Russia British infantry found themselves more seriously involved. Another early arrival was the 2nd/10th Royal Scots. They had been a cyclist battalion in Ireland when they were ordered to sail to Archangel, the principal Russian port in the White Sea. Disembarking on 25 August, 1918, they were immediately instructed to move up the River Dvina. Brushing Red detachments out of the way, short of food, and often under heavy rain, they advanced some 150 miles, in company with the US 339th Infantry Regiment, before being forced to halt by a superior Bolshevik force. In late November two battalions of Green Howards and the 11th Royal Sussex landed at Murmansk. In spite of successfully enduring the bitter winter, increasing Red pressure and growing disaffection among the often British-led White Russian units caused the British Government to decide in April, 1919, that British troops should be withdrawn from North Russia before winter closed the ports once more. This had to be kept secret from the White Russians for fear of increasing their rebelliousness. In order to cover the withdrawal a brigade of volunteers was hastily raised and sent out from Britain. It consisted of Nos 1 Special Companies of the King's Royal Rifle Corps and Middlesex Regiment, largely formed from officers and men who had seen extensive service on the Western Front, 45th and 46th Royal Fusiliers, and the 2nd Hampshires. They, too, found themselves involved in heavy fighting in the course of which Sergeant Samuel Pearse of the 45th Royal Fusiliers won a posthumous Victoria Cross for single-handedly charging and clearing a Red blockhouse which had been causing trouble during an attack. Corporal Arthur Sullivan of the same battalion was also awarded the Victoria Cross in a separate action. Eventually, in September, the entire British contingent was withdrawn.

In contrast, in South Russia, the other area of Allied intervention, which was in support of the White generals Denikin and Wrangel, no British infantry took part, although a tank detachment and the RAF did. Many of those who fought in Russia were bewildered, bemused, and quickly disillusioned, especially by the growing efficiency of the Reds and the poor quality of the White soldiers, but at least one professional soldier, CSM Fred Neesam of the Green Howards, did not begrudge his service there: 'I enjoyed every minute of it, despite the fact that the going was hard.'

Another immediate postwar problem, Ireland, also proved to be a frustrating experience. True, the British Army, although caught by surprise, had had little difficulty in putting down the Easter Rising of 1916, which, in any event did not enjoy popular support in Ireland. The Irish Nationalists were, however, able to turn this to their own advantage, portraying those who had been executed or imprisoned as martyrs in the cause of an independent Ireland and began to bring the population in the south increasingly behind them. This was helped, even though it was quickly dropped, by a British Government proposal in early 1918 to extend conscription to Ireland. The net result was that Sinn Fein won 75 per cent of the Irish seats in the Khaki Election of December, 1918. When they saw that this was not sufficient for the British to release their hold on the country, Sinn Fein unilaterally declared independence and set up its own parliament, the Dail. Some more impatient spirits now believed that the only way forward was to provoke an open revolt throughout the country, and in January, 1919, murdered two members of the Royal Irish Constabulary (RIC) as the Dail sat for the first time. It was the RIC which initially bore the brunt, so much so that by the beginning of 1920 it had to be strengthened by volunteers from the mainland, the Black and Tans and the Auxiliaries, largely made up of ex-soldiers who had found it difficult to settle down in civilian life after the end of the Great War. It was they who now spearheaded efforts to snuff out the Irish Republican Army, but very quickly their methods of doing so degenerated into a lawlessness akin to that of their enemy.

During the first half of 1919 the British Army was very much in a support role, providing escorts for the RIC, enforcing curfews, searching houses, and guarding government buildings. Private Swindlehurst of The Lancashire Fusiliers recorded how, before carrying out a search, the troops would report to the local police station, be briefed, and then set off accompanied by a constable. The searches themselves were not improved by the 'stale air', largely caused by 'the fumes from little red-globed oil lamps burning before a highly coloured picture of the Blessed Virgin Mary', with 'the windows tightly shut'. Often the troops were spat upon and taunted, but matters became worse in the second half of 1919 when the IRA began to attack the Army as well. Hit and run tactics in the form of flying columns or simply the danger of being shot down in the streets by a 'civilian' were totally alien to soldiers who had just returned from the Great War and the desire to run amok and blindly seek revenge when a comrade was killed was very strong. Yet, the battalions, with a few exceptions, did manage to maintain their discipline and, indeed, generally regarded the savage reprisals wrought by the Black and Tans and Auxies with distaste. Eventually it became clear to the British Government that military means could not enforce a solution to the problem and, after months of negotiation, the Irish Treaty of December, 1921, brought about the creation of the Irish Free State out of the Twenty-Six Counties, leaving just the six predominantly Protestant counties of the north-east tied to the mainland. This enabled the Army to withdraw thankfully. Few soldiers stationed in southern Ireland at that time would have thought that fifty years later their grandchildren in uniform would be experiencing the same all over again, albeit in the north of 'John Bull's Other Island'.

1919 also saw the British Army in action elsewhere in the world. Part of this arose from Britain being granted the mandates of Palestine, Transjordan and Mesopotamia, former territories of the Ottoman Empire, through the Versailles Peace Treaty. In northern Mesopotamia the Kurds had hoped that the British would allow them to realize their dream of an independent Kurdistan, and it took two Indian infantry brigades, supported by cavalry, armoured cars and aircraft to subdue them when they became aware that this was not the British intention. In the south there was also unrest among the Marsh Arabs. The following year matters were aggravated by increasing nationalism, which turned to open revolt, among the Muslim majority in the country and because the garrison had been heavily reduced. The 3rd Manchesters suffered significant casualties as part of a relief column which was ambushed, with 180 killed and 160 captured. The latter, however, were well treated and were released after a few weeks. Everywhere the British were forced on to the defensive and it was only after reinforcements had been sent from India that relative peace was restored. The creation of the kingdoms of Iraq and Transjordan also helped pacification.

In 1922, however, a new policy was introduced. This was the brainchild of Hugh Trenchard, Chief of the Air Staff, who saw it as a means of safeguarding the independence of the Royal Air Force. Air control relied on airpower to keep the tribes in order and demonstrated that the RAF, supported by its own armoured cars and locally raised levies, could do this with many less men than the Army. The latter may not have liked it: the Chief of the General Staff, Sir Henry Wilson, saw it as a threat to the Army's traditional role of imperial policing, but at the time military commitments elsewhere meant that the Army was desperately overstretched.

There was also trouble with Afghanistan, triggered by a wave of agitation for more self-government which swept northern India early in 1919 and which culminated on 13 April at Amritsar, when Brigadier General Dyer ordered Gurkha troops to fire on a hostile crowd. The new King Amanullah of Afghanistan, who despeately needed something to divert domestic threats to his throne, seized on Amritsar as manna from heaven. He proclaimed a holy war against the British, and at the beginning of May his troops occupied a disputed locality called Bagh in the Khyber Pass. This took the army in India by surprise. Most of the Regular Indian Army formations were still in the Middle East and there was little left in the country besides 'green' Indian troops and disaffected Territorial battalions, who only wanted to return to Britain. Even so, a column was hastily formed and within a week had driven the small Afghan detachment from Bagh. It was then ordered to advance by stages to Kabul, and, after a stiff fight at Dakka five miles west of Bagh, set off for Jellalabad, thirty miles further west.By coincidence the column included a Territorial battalion of The Somerset Light Infantry, whose forebears, the 13th Foot, had won the nickname 'The Illustrious Garrison' for their defence of Jellalabad during the First Afghan War (see Vol I). The Somerset Territorials, who had already earned for their regiment the battle honour 'NW Frontier India, 1915', saw little romance in the fact that they wore 'Jellalabad' as part of their cap badge and were only too happy when a halt to the advance was now called. The reason for this was that a sizeable Afghan force had crossed the frontier elsewhere and very quickly overran all the outlying British posts in Waziristan. Another column was organized, this time under Dyer, and moved to confront the Afghans. At this moment King Amanullah got cold feet and ordered his troops to cease hostilities. This was just as Dyer's artillery had begun to knock out the Afghan guns. The Afghans hastily withdrew back across the border, harassed by armoured cars and aircraft, and the war, if it could be called that, ended as suddenly as it had started, after just twenty-nine days. Now, finally, the British Territorials could pack up and return home.

The Irish Troubles had made life difficult for the five southern Irish regiments – Royal Irish Regiment, Connaught Rangers, Leinster Regiment, Royal Munster Fusiliers and Royal Dublin Fusiliers. True, none of them were actively involved in the operations, but reports by the newspapers, often exaggerated, of the atrocities perpetrated by the Black and Tans did put the soldiers, worried about their families, under severe pressure. Yet, apart from one instance, they remained faithful to the Colours. The exception was the 1st Connaught Rangers who were stationed in India at Jullundur. At the end of June, 1920, a group of NCOs and men staged a 'sit in' in the guardroom in protest at the Black and Tans. They refused to do duty and were arrested and placed under guard in another nearby camp. Private James Daly led some other men in a sympathy march to the Officers' Mess and informed his company commander that they, too, would not do duty until British 'oppression' in Ireland had ceased. After being warned of the gravity of their actions, the men hoisted an Irish flag over their bungalow, but were persuaded by the Padre to hand over their rifles, although they were allowed to retain their sidearms. But then rumours began to circulate that the original 'mutineers' had been attacked and murdered by English machine guns. Daly and others became thoroughly worked up and, armed with their bayonets, stormed the battalion armoury. The guard was forced to open fire, killing two and wounding another, before the remainder surrendered. In all, 75 men subsequently were court-martialled, with fourteen being acquitted. Of the remainder all, apart from Daly, were sentenced to varying terms of imprisonment ranging from life for eight who were originally awarded death sentences, but commuted on review by the Commander-in-Chief, down to one year. In the event, none served more than three years in gaol. Daly himself was executed by a firing party from The Royal Fusiliers on 2 November, 1920. It was a sad occurrence for a very famous regiment, but less than ten per cent of the battalion were involved and the remainder continued to do their duty.

In 1922 came the so-called 'Geddes Axe', when Prime Minister Lloyd George appointed a committee under Sir Edward Geddes to drastically reduce public expenditure, defence included. The upshot was that the Army had to be reduced by 50,000 men. By now, with the commitments in Russia and Ireland finished, much of the Middle East handed over to the RAF, and India comparatively quiet, the moment appeared to have arrived when this could be done. It was the cavalry who suffered worst, with sixteen regiments being amalgamated, but the infantry had to take its share. The Royal Fusiliers, Worcesters, Middlesex, 60th Rifles, and Rifle Brigade lost their third and fourth battalions and, in the context of the situation in Ireland, it was decided that the Southern Irish regiments and The Royal Irish Fusiliers should be disbanded. The last-named was saved through the personal intervention of King George V, on the grounds that it recruited throughout Ireland, but was reduced to one Regular battalion and had to share a depot at Omagh with The Royal Inniskilling Fusiliers, who were also reduced to a single line battalion. There was no saving of the other five and in July, 1922, they were formally disbanded. They surrendered their colours to King George V for safekeeping in St George's Hall, Windsor. His address on this occasion reflected much of the meaning of the Colours to a regiment:

'We are here today in circumstances which cannot fail to strike a note of sadness in our hearts. No regiment parts with its Colours without feelings of sorrow. A knight in days gone by bore on his shield his coat-of-arms, token of valour and worth; only to death did he surrender them. Your Colours are the record of valorous deeds in war of the glorious traditions thereby created. You are called upon to part with them today for reasons beyond your control and resistance. By you and your predecessors these Colours have been referenced and guarded as a sacred trust – which trust you now confide in me.

As your King I am proud to accept this trust, but I fully realize with what grief you relinquish these dearly prized emblems; and I pledge my word that within these ancient and historic walls your Colours will be treasured, honoured, and protected as hallowed memorials of the glorious deeds of brave and loyal regiments.'

Yet, for many years, until 1949, when Eire became a republic and left the British Commonwealth, the five regiments continued to appear in the Army List, their battle honours gained over, in some cases, nearly 250 years of service, given in full.

This restructuring also brought about amendment to many regimental titles. Thus, for political reasons, The Royal Irish Rifles became The Royal Ulster Rifles, but in other cases regiments sought to improve on or correct anomalies in the 1881 titles. The Princess of Wales's Own became The Green Howards, by which they had been more commonly known for some two hundred years, while The Royal Welsh Fusiliers reverted to The Royal Welch Fusiliers. The *Daily Herald* commented 'Strewth' on the latter, but 'the spelling with a *c* was,' as the poet Robert Graves, wartime officer in the Regiment, wrote, 'as important to us as the miniature capbadge worn at the back of the cap was to the Gloucester ... "Welch" referred us somehow to the archaic North Wales of Henry Tudor and Owen Glendower and Lord Herbert of Cherbury, the founder of our regiment; it dissociated us from the modern North Wales of chapels, Liberalism, the diary and drapery business, slate mines, and the tourist trade.'

Other changes included the ending of the traditional Militia concept, but rather than disband the Militia battalions, they were merely allowed to waste away. The Territorial Force was, however, retained under the new title of the Territorial Army, although it did suffer from some reductions, The London Regiment, for instance, being reduced by three battalions, although one of these was converted to become part of the newly formed Royal Corps of Signals. The Territorial Divisional structure was also kept in the form of the 42nd–44th and 46th–56th Divisions.

The organization of the British Army was much influenced by what was called 'The Ten Year Rule'. This had been drawn up by the Government on the premise that Britain would not be involved in another major war for ten years. It was renewed every year until 1932, when the period was reduced to five years. Consequently, priority for defence once more rested with the Empire, something which was welcomed by many soldiers who had served pre-1914 and who had not considered trench warfare to be 'real soldiering'. The Haldane system of one line battalion serving overseas while the other provided drafts for it at home was resurrected.

A new organization for the infantry battalion was, however, introduced in 1920. Battalion headquarters now had nine officers and 109 soldiers, and included a Headquarter Wing, which controlled all the administrative elements. The four rifle companies each had six officers and 209 soldiers, with platoons retaining two rifle and two Lewis Gun sections, close to the wartime structure. In 1922, on the disbandment of the Machine Gun Corps, the battalion was augmented by a machine-gun platoon, with eight Vickers, and this became part of the Headquarter Wing. In practice, though, the battalions at home in Britain were often little more than skeletons. On manoeuvres, infantry sections had to be represented by flags, with football rattles being made to represent machine-gun fire. These frustrations were often overcome by devoting much time to sport, which served to maintain the battalion's spirit.

Indeed, an atomosphere of stagnancy existed except in one quarter, the realm of armoured warfare. Here, two visionaries, Captain Basil Liddell Hart and Colonel (later Major-General) J.F.C.Fuller, preached the gospel of the tank as the weapon of the future. Sufficient official note was taken of them for an experimental mechanized force to be set up on Salisbury Plain for the 1927 training season. It consisted of two battalions of the Royal Tank Corps, five artillery batteries, an engineer company, and the 2nd Somerset Light Infantry, which was organized as a three-company machine-gune battalion mounted in half-tracks and lorries. In addition, for much of the time the 2nd Cheshires were attached to the Force as a lorry-borne infantry battalion. During the exercises the Force was pitted against 3rd Infantry Division, whose infantry had a hard time of it. In the words of the Experimental Force's commander, 'Whatever distances they covered on foot, the Mechanized Force could still apparently "make rings round them".' Morale suffered, and General Jock Burnett-Stuart, commanding

3rd Division, tried to set the record straight by stating that the Force was sensitive to terrain and vulnerable to small arms fire since only some of its vehicles were armoured. Fears, though, that the infantry might find themselves turned into 'tank marines', to use Fuller's term, or mere 'fortress troops', were quickly laid to rest by the Chief of the Imperial General Staff, Sir George Milne, who made it clear that financial restrictions meant that the development of armoured forces could proceed only very slowly. Even so, each infantry battalion did receive three Carden-Lloyd tracked carriers, equipped with Hotchkiss light machine guns for reconnaissance purposes. The machine-gun establishment was also enhanced by an additional four guns, with a further four to be provided on mobilization. The machine guns were therefore formed into a separate company made up of four-gun platoons, but because of the shortage of manpower, the crews for the third active platton had to be found from the rifle companies.

One reason why the British Government was happy to hand over the Middle East to the RAF was triggered by the Chanak Incident of 1922. Turkey's new ultra-nationalist leader, Mustapha Kemal, refused to accept one of the Allies peace terms, namely that the Greeks, traditional enemy of the Turks, should occupy half of Anatolia on the Turkish mainland. Fighting broke out between the Greeks and Turks, with the latter quickly gaining the upper hand. The wartime Allies, less the USA, which had now reverted to isolationism, decided that they must act with a show of force. The British despatched ships and a contingent of troops, which included three battalions of Foot Guards (Grenadiers, Coldstream and Irish). First to arrive and be landed at Chanak on the Asiatic side of the Dardanelles, in May, 1922, were the Irish Guards, with the future Field Marshal Viscount Alexander in command. During the initial few weeks, while the Irish Guards were becoming acclimatized, one of Alexander's officers, Francis Law wrote:

'We were warned of the risk of drinking a fierce local absinthe, for it had a treacherous delayed action, our men called it "Doosico". One morning a number of them who had returned to barracks the night before "clean and properly dressed" were found to be drunk after drinking their breakfast tea. They were dealt with, but not too harshly, for they had by demonstration done everyone a service.'

Initially relations with the Turks were good, and the Irish Guards were able to see the sights and honour the official King's Brithday in the traditional way by trooping their Colours. But, in late summer, the situation became more tense and it seemed for a time that hostilities would have to be resumed against Turkey. Gradually, through much diplomacy, the climate thawed, but not before the Trooping of the Colour had been performed once more at Chanak, this time with all three Guards battalions. With the signing of the Treaty of Lausanne in July, 1923, which formally ended the Great War as far as Turkey was concerned the British troops were able to return home.

Three years later came the next challenge, and even further afield. China had long been torn by internal feuding and much of it was ruled by the warlords, who merely bickered among themselves. But there was also growing resentment among many thinking Chinese over the way that Western countries were milking their economy, and nowhere was this better symbolized than in the International Settlement at Shanghai. In 1925 Chiang Kai-shek set out from southern China determined to overcome the warlords and bind China into one country. By the end of 1926 he was approaching Shanghai and it seemed as though he was bent on getting rid of the International Settlement. Consequently two brigades were sent from Britain and one from India. The former, however, had to be brought up to strength with reservists. They spent the whole of 1927 in Shanghai, but were not actually called upon to fight the Chinese, since Chiang Kai-shek, who entered Shanghai in March, modified his attitude to the 'foreign devils'. This was mainly because he needed money to maintain his forces and he could only obtain this through the merchant class, who, in turn, were closely bound to the foreign concessions in the country. Furthermore, he discovered that the Soviet Communists were using him for their own ends. Even so, from then on two infantry battalions were stationed in Shanghai, although they had to be withdrawn in August, 1940, in the face of Japanese political pressure.

India, though, continued to be the main overseas posting for the infantry, with approximately forty-five British battalions being stationed there at any one time during the 1920s, as compared to some sixty at home. The tour of a battalion abroad was much the same length as it had been prior to the Great War. Thus, the 2nd Seaforths were sent to India in 1919 after spending the war on the Western Front. They returned to Britain in 1934. They were initially stationed at Meerut, where the Mutiny had started in 1857, and then did a tour on the Frontier at Landi Kotal and Nowshera, before settling down in Jhansi, south of Agra. In 1930 they returned to the frontier once more for operations against the Afridis. This earned those who took part the 1908 Indian General Service Medal with the 'North West Frontier, 1930–31' bar. Leaving India in 1932, they spent two years in Palestine before finally returning home.

Frontier warfare had changed little. The concept of the punitive column was still used, only now aircraft, armoured cars and even light tanks were used to support it. Nonetheless, the tribesmen continued to be as wily as ever. Away from the Frontier the British infantryman increasingly found himself engaged in military aid to the civil power. Rising nationalism brought about civil unrest, which he would often be called upon to quell. After the 1919 Amritsar 'massacre' the regulations of how rioting should be controlled had been considerably tightened up and the soldier had to be prepared to endure insults and brickbats without reacting or flinching in any way, a severe test of discipline.

Even so, the quality of life had improved since pre-1914. Electricity had come to the cantonments, as had the cinema. Married soldiers had more opportunity to bring their families out to India, but the single soldier often spent as long as five years in the 'Shiny' without social contact with white girls, and banning of military brothels in the 1890s (see Vol I) had forced him into the native quarter, with its high risk of venereal disease, for sexual gratification. To keep this to a minimum much emphasis was placed on sport on the principle of *mens sana in corpore sano*. In contrast, in Alexandria in the early 1930s there was a garrison brothel, which the commanding officer of the 1st Royal Warwicks, B.L. Montgomery, actively encouraged his men to use, with his medical officer being responsible for inspecting the girls. Even so, life overseas could be still very monotonous, as a member of the 1st Cameronians, who spent most of the inter-war years in China and India, records:

'You lived between the barrack-room and the canteen, without any social life at all. For all the years I was in India . . . I was never in a [private] house. . . .

It's strange to look back on some of it now. There was a ritual every evening. The men would make themselves absolutely spotless – uniform pressed, boots polished, hair plastered down, bonnet on just so – just as if every one of them had a girl-friend waiting at the gate. They went straight down to the wet canteen and got drunk. That was what they got dressed up for.

It didn't matter what continent you were in, it was always the same. Everything was organized within the battalion – football, rugby, hockey, boxing. You got the odd leave in a rest camp, but that was just the army again without the parades. On 3 1/2d [1.5p] a day ration money you couldn't spend a leave anywhere else, without going broke as soon as you started. You got paid home leave once in six years before you left.'

One of the more unusual tasks that came the infantry's way occurred in early 1934. This was to assist in supervising the League of Nations plebiscite on the Saarland. The 13th Infantry Brigade, based at Catterick in Yorkshire, was earmarked and spent three months ensuring that the plebiscite, which resulted in an overwhelming majority of the population wanting to be reincorporated into Hitler's Germany, was carried out without violence or other irregularities.

A more demanding challenge came in 1936 with trouble in Palestine. Resentful of the increasing number of Jewish immigrants, the Arabs came out in open revolt against British authority through the medium of guerrilla warfare. Police stations were attacked, roads mined, telegraph lines and railways sabotaged and ambushes mounted. At the time the garrison in Palestine consisted of an RAF armoured car company, the 900 men of the Transjordan Frontier Force, two bomber squadrons and two infantry battalions – 1st Loyals and 2nd Cameron Highlanders. Reinforcements were initially sent across the Suez Canal from Egypt. As it happened, the garrison here had been swelled the previous year because of the possibility of war with Italy over the invasion of Abyssinia. During the summer of 1936 clashes with the Arabs were frequent. Typical of them was one involving two Rolls-Royce armoured cars of the 11th Hussars and a platoon of Loyals. Searching in the Samarian hills for fugitives of a clash the previous evening they came across Arabs setting up a roadblock just inside the entrance to a narrow defile. They brushed these aside and continued under fire through the defile to a village the far end, which was searched, without result, by the Loyals. On their way back through the defile they were ambushed by a band of sixty Arabs. The platoon commander was wounded, but, in spite of this and their numerical inferiority, the Loyals debussed from their truck and, supported by the machine guns of the armoured cars, charged the Arabs. The cars also used their radios to call up air support, which arrived in the shape of two aircraft with bombs. The result was twenty-six guerrillas killed or captured, with the remainder only escaping because the Loyals were so few in number.

In autumn 1936 further troops were sent from Britain and the force in Palestine quickly swelled to two divisions, one in the north and the other the south. This was still not enough to quash the rebellion, largely because Whitehall refused to condone the implementation of martial law, preferring the 'velvet glove' approach.

Not until after the Munich Agreement of September, 1938, when the threat of war in Europe appeared to have receded, did the British Government eventually relent, and the situation was then brought under control. In the meantime the main offensive tactic employed was the cordon and search. This involved swooping on a village suspected of harbouring guerrillas, surrounding it so that no one could escape, and then carrying out a detailed search. Often, though, the village dogs would give the show away, allowing suspects to melt away. It was a frustrating business, but did, in conjunction with two major campaigns on the Indian North-West Frontier, help to prepare the infantry for the major war which became increasingly inevitable.

It is significant that the two divisional commanders in Palestine in 1938, Dick O'Connor and Bernard Montgomery, were both infantrymen who had served together a few years before in Egypt. O'Connor had been a company commander in the 1st Cameronians and Montgomery Commanding Officer 1st Royal Warwicks. Both would later play leading roles in the campaign against the Axis in North Africa.

British defence preparedness reached its nadir at the same time as Hitler came to power in Germany. When the Government did eventually begin to recognize the growing threat to peace in Europe and began to rearm the Army was placed a poor third behind the other two services. Not until Leslie Hore-Belisha became War Minister in 1937 was it given the opportunity to modernize.

The Army as a whole welcomed his reforms. The soldier's pay, which had been frozen since 1925, was increased in order to attract desperately needed recruits to make good a shortfall of 20,000 men. Hore-Belisha set in train a barrack-building programme to give the soldier a more civilized environment than the forbidding Victorian barracks and Great War wooden huts in which the Army at home lived. He instituted a supper meal for the soldiers, even though it was little more than bread and jam and cocoa, and allowed them a small butter ration rather than just margarine. He also unblocked promotion for junior officers, many of whom had served as subalterns for fifteen years or more. Indeed, it was this stagnancy, together with the pacificism that had influenced the intelligentsia during the last Twenties and early Thirties, which had dissuaded many from a military career. Nevertheless, it was not a situation which could be improved overnight. Hore-Belisha's immediate solution to the shortage of junior officers, especially in the infantry, was therefore to create a new grade of Warrant Officer, the Warrant Officer Class III or Platoon Sergeant Major, who would act as platoon commander, and the first batch of 1000 senior NCOs were appointed to this rank in October, 1938. It did not, however, prove to be a popular measure. The WO3 was regarded as neither 'fish nor fowl' and the rank was abolished early in the Second World War, once sufficient officers had been trained.

As the clouds of war loomed larger over the horizon the infantry underwent a number of changes. In 1936 its support weaponry was enchanced by the addition of three 3-inch mortars; these and the machine guns now formed the Support Company. The following year a new battalion establishment was agreed. The Headquarter Wing was replaced by a Headquarter Company, which had a signals (lamp, flag, heliograph, line) platoon, an anti-aircraft platoon, with light machine guns mounted on trucks, a mortar platoon (now just two 3-inch mortars), and an administrative platoon. There were now four rifle companies, each still with four platoons, but these were now reduced to three sections, each equipped with a new light machine gun, the Bren, which had originally been designed by the Czechs and was to give the British Army sterling service over many years. The platoon also had a 2-inch mortar and a Boys anti-tank rifle, which fired a 0.55 inch steel-cored bullet capable of penetrating 20mm of armour at 500 yards. One significant result of the three-section platoon was that infantry drill was now based on three ranks rather than four.

During 1938 and 1939 the battalion was further strengthened by the addition of a carrier platoon, with ten bren-gun carriers for reconnaissance, and a pioneer platoon. It also had its transport totally motorized, in contrast to the German infantry battalion, which still retained a high proportion of horse-drawn vehicles. These improvements only applied to battalions based at home, and excluded those which had been sent to Palestine. Indeed, Montgomery wrote as late as July, 1939, that the troops in Palestine were 'experts in fighting a savage and mobile enemy, who is armed only with the rifle and knife – they know nothing of other forms of war. . . . Such things as Pioneer platoons, Carrier platoons, AA platoons, etc. are unknown – the situation regarding specialists is very bad.'

The new infantry battalion lacked medium machine guns, and the reason for this was a decision to concentrate them at divisional level as machine-gun battalions, thus providing a belated admission that the disbandment of the Machine Gun Corps may have been a mistake. The original plan was for twenty-eight battalions, including two of Foot Guards, to be converted to this role. In the end, merely the Line and some territorial battalions of the Northumberland Fusiliers, Cheshire, Middlesex, Manchester, and London Regiments underwent conversion. Again, though, Palestine inter-fered with this, the 1st Manchesters having only just received their machine guns before they had to put them in store and take ship for the Middle East.

Steps were also taken to expand the Army as a whole. The infantry's share in this was originally planned to be fourteen additional line battalions, but, because of shortages in equipment and manpower, only four were raised. The Royal Irish Fusiliers and The Royal Inniskilling Fusiliers were each given back their second battalions in 1937, and two years later the Irish and Welsh Guards each formed a second battalion. Then, at the end of March, 1939, Hore-Belisha ordered a doubling in the size of the Territorial Army.

As far as the Territorial infantry were concerned, the groundwork for this had been laid at the time of Munich, with each existing battalion creating a 'shadow' battalion, which would only come into being on mobilization. This was just as well, since there were not the volunteers to man them, and in May, 1939, Hore Belisha had to resort to limited conscription. All 20-year-olds were called up for six months' service with the Regular Army as 'Militiamen' prior to joining new second line Territorial Divisions, which would duplicate existing ones as follows:

Second Line	First Line
9th (Scottish)	51st (Highland)
12th	44th (Home Counties)
15th (Scottish)	52nd (Lowland)
18th	54th (East Anglian)
23rd	50th (Northumbrian)
38th (Welsh)	53rd (Welsh)
45th (Wessex)	43rd (Wessex)
59th	46th (North Midland)
61st	48th (South Midland)
66th	42nd (East Lancashire)

It should, however, be pointed out that in the mid-1930s, in view of the air threat to Britain, a number of TA battalions had been converted to the anti-aircraft role, both manning guns and operating searchlights. These were respectively rebadged Royal Artillery and Royal Engineers. Indeed, the whole of 46th Division was converted to this role, before being reformed as an infantry division in 1939. Others, too, became Royal Signals regiments, and seven were converted to tanks as the 40th–46th Royal Tank Regiments. This largely explains why not every first line Territorial division had a second line equivalent.

The final major change prior to the outbreak of war in September, 1939, was to field uniform. The brass-buttoned khaki tunic, slacks, and long puttes, which had clothed the soldier for the past thirty years, finally gave way to something more modern in concept. This was Battle Dress, designed on the lines of a ski suit, at least in theory, and worn with anklets rather than puttees. While it was quicker to put on, and had more pockets, it was not as smart as the old uniform. The short jacket, although secured to the waistband of the trousers by two buttons at the rear, also tended to ride up, thus exposing the kidney region to chills. A khaki sidehat replaced the peaked hat, but also proved unpopular, being difficult to keep on the head, especially in a wind. New webbing equipment was also introduced. Simpler to put together than the old, the most significant difference was the substitution of eight small ammunition pouches by two large ones in which Bren gun magazines could be carried. Yet the infantryman of 1939 still retained the same rifle and bayonet with which his predecessor had gone to war in 1914 and continued to wear the 1914–18 helmet.

CHAPTER FOUR

The First Shock of Battle (1)
NW Europe 1939–1942

T HE British Army in September, 1939, had five Regular infantry divisions (1st-5th) based at home, and a sixth (7th) in Palestine.* Each was organized in much the same way as those of 1918, with three brigades, each of three battalions. Divisional troops consisted of a cavalry regiment equipped with light tanks, three field regiments (with 25-pounder gun/howitzers) and an anti-tank regiment Royal Artillery (2-pounder anti-tank guns), three field companies and a field park company Royal Engineers, and a machine-gun battalion. These divisions were immediately brought up to strength with reservists, who made up about 50 per cent of the establishment, an indication of how weak battalions were prior to mobilization. In addition, there were thirteen first line Territorial divisions in being, one first line (46th) being reformed, and ten second line being formed. All needed training and properly equipping before they could be considered ready for war.

Under the terms of an agreement made with the French in April, 1939, the BEF began to cross to France, as it had in 1914, towards the end of September. The first wave consisted of 1st-4th Infantry divisions, organized in two corps, and was to take up position on the left of the French armies and opposite the Belgian frontier. The days before embarkation were spent in a mad scurry to make good equipment deficiencies, an indication of how late preparation for war had been left, and the divisions were still especially short of ammunition, anti-tank and anti-aircraft weapons. It was thus just as well that there would be no immediate encounter battle, as there had been in 1914. Instead the BEF was to find itself enduring a long wait for 'the balloon to go up', and during this time priority was given to the construction of defensive works. In most formations this was to the exclusion of almost all else. A subaltern in the 1st Border Regiment:

'So much of the time was spent in digging or drinking that there was

---

* The other division in Palestine, 6th, had been broken up earlier in the year.

little left for training. During those eight months I don't think I took part in one field exercise, though I did construct a railway-station yard, build a road and turn a stream into an anti-tank obstacle. No, I'm wrong; not a complete obstacle. When it was half-finished we left it to build the road and left it to construct the railway yard. That we did finish, but a late frost almost immediately undid our work.'

Even so, morale remained remarkably high. The same subaltern could not recall more than one court-martial throughout the time his battalion was in France. He also noted that among the reservists in his platoon were 'two first-class aircraft fitter-mechanics, a post office linesman and a tradesman carpenter' and that 'it must have come as a shock to some of them to go straight from "civvie street" to the front line, with no intermediate period of preparation; but they took it like men, in good heart and with a ready laugh.' Later on those with technical skills would be largely weeded out of the infantry and transferred to other arms and services in which their trades could be more directly applied.

Part of the reason for the lack of tactical training was the French insistence that radio silence must be preserved at all times. One division, however, was determined not to let the grass grow under its feet. Once the Allied plan for an advance into Belgium in the event of a German attack had been agreed with the French in November, Montgomery began to put his 3rd Division through a series of movement exercises, including taking up a defensive position, by day and by night. More significantly, he made his troops also practise withdrawals. This was training which would stand the Division in good stead in May, 1940. Further valuable experience was achieved through sending brigades for attachments to the French in the Maginot Line. They were given a sector opposite the German frontier south-east of Luxembourg and occupied positions forward of the Maginot fortifications themselves. Here fighting patrols, often clad in white oversuits to camouflage themselves against the snow, traversed no-man's-land, hoping for a clash with German patrols. There were a number of successes, but also casualties. Indeed, it was during one of these patrol actions, on 9 December, 1939, that Corporal Thomas Priday of the 1st King's Shropshire Light Infantry was fatally shot, the first man in the BEF to be killed in action.

One of the favourite patrol weapons was the Thompson machine gun, so beloved of American gangsters, some of which had been hastily imported. Certainly it could produce a high volume of fire, but it did have its drawbacks. After one successful clash, the patrol in question had had to drag back the body of a German soldier some three miles through snow-covered fields for intelligence gathering purposes. The corpse was laid out in the battalion aid post, where 'in amazement we found that although the unfortunate man had been hit by a positive fist-full of .45 Tommy Gun bullets, only one had killed him – all the others were lodged in his clothing and the various straps and webbing of his equipment. This was a shock and momentarily impaired our faith in the beautiful new sub-machine guns which had only so recently come our way.'

This long period of waiting did enable the BEF to be gradually strengthened. In December 5th Division arrived in France, to be followed in January by three first line Territorial divisions, 48th, 50th, and 51st. Finally, in April, 1940, came 42nd, 44th and 46th Divisions, and two second line Territorial divisions. These last named, 12th and 23rd, together with the 46th Division, were sent initially to carry out labouring tasks on the lines of communication, the idea being that they would complete their training in France. They also lacked artillery and many of their administrative elements. They would shortly pay dearly for this.

Back in Britain, on the day that war was declared Parliament had passed the National Service (Armed Forces) Act, which made all males between the ages of 18 and 41, other than those in reserved occupations, liable to conscription. There would, however, unlike in 1914, be no mass flocking to the Colours. The first priority was to complete the training of the Militiamen and the equipping of the enlarged Territorial Army. Only then would further men be called up as and when the equipment was available for them. Likewise there was no question of immediately withdrawing Regular battalions from overseas and replacing them with Territorials. They would remain where they were.

It was not, however, just the defence of France and the Low Countries which occupied the minds of the planners during the Phoney War. At the end of November 1939, the Soviet Union invaded Finland. Surprisingly, the Finns put up a fierce resistance, and the British and French began to think in terms of sending a force to their aid. British preparations included the formation of a special ski battalion, 5th Battalion Scots Guards, which was sent off to Chamonix in the French Alps to perfect its techniques. Disbanded after the fall of France without seeing any active service, a number of its members later made names for themselves when serving with Special Forces. Notable among them was David Stirling, the founder of the Special Air Service.

In March, 1940, the Russians finally overwhelmed the Mannerheim Line and forced the Finns to cede for peace. Allied attention therefore switched to neutral Norway, through which Swedish iron-ore, vital to Germany's war effort, passed. By mid-March plans for sending troops to Norway had been finalized.

Matters, though, were accelerated by the German invasion of Denmark and Norway on 9 April. Priority, of course, lay with the defence of France, and the troops initially earmarked for what was originally expected to be an unopposed landing in a friendly country were largely made up of the 49th (West Riding) Division TA and the Regular 24 Guards Brigade. In the event, three landings were made on the Norwegian coast, but such was the initial confusion that battalions embarked without half their stores and equipment. It was 148 Infantry Brigade which suffered worst. Originally it was to make an unopposed landing at Stavanger and boarded two cruisers for this purpose. This order was then countermanded, and it was hastily put ashore, losing some of its equipment in the process. It was now told that its objective was Namsos, and re-embarked on a transport. Three days later orders came through that it was to be taken at once in five warships to Andalsnes. Everything therefore had to be reloaded, and more equipment was lost. The result was that it eventually landed with no mortar ammunition, lacking most of its communications equipment, without range-finders for its anti-aircraft guns, and precisely one truck and three motor-cycles as transport. Half of one battalion also had to be left behind because of lack of room. It had distinct echoes of those disastrous amphibious expeditions during the early part of the Napoleonic Wars, and it is hardly surprising that the brigade, especially with its still half-trained troops, was unable to make much impression on the Germans.

In contrast, 15 Infantry Brigade (part of 5th Division), which was hastily sent from France and also landed at Aandalsnes to support 148 Brigade, performed well. It was to the 1st King's Own Yorkshire Light Infantry of this brigade that the distinction fell of being the first British Regular troops to openly engage the Germans in the field during the Second World War. Supported by eight French 25mm Hotchkiss anti-tank guns, the Koylies successfully held German troops supported by tanks for 24 hours at Kvam over 25–26 April. Then, lacking virtually any artillery support and under total enemy domination of the air, the Brigade, which also included the 1st Green Howards and 1st York and Lancaster, fought a dogged withdrawal, inflicting many casualties on the enemy, back to Aandalsnes and enabled the port to be safely evacuated. This, however, was one of the very few bright spots in an otherwise disastrous campaign, although it would not be until early June when the last Allied troops left the country. These consisted of two French demi-brigades, a Polish brigade, and 24 Guards Brigade, which had just forced the surrender of the German garrison at Narvik. In the meantime the Allies had experienced even greater disaster elsewhere.

On 10 May, 1940, the long-awaited German attack on the West was finally launched. The BEF accordingly crossed the Belgian frontier and by nightfall had taken up position along the River Dyle. On the 14th the Germans closed to the river and began to attack in earnest on the following day. The 2nd Royal Ulster Rifles were holding the town of Louvain and soon found themselves embroiled in a stiff fight near the railway station. One of their platoon commanders:

'The enemy now got a Spandau [machine gun] across the line between the Rifles and the 1st Grenadiers on the left. This opened up on the platoon from the rear, the bullets striking the wall at the top of the ladder and the parados of the trench. About the same time they worked a heavy 20mm machine-gun forward and this opened up on the position from about 100 yds range, the heavy armour-piercing bullets knocking the parapet back into the trench and filling the air with the cinders and earth of which it was composed. The shouts of the German NCOs, accompanied by the blowing of whistles, could now be heard above the din as they formed up to rush the position. A line of railway trucks on their side of the embankment gave them shelter. Cpl Gibbens had now taken over a Bren, and to get a better aim at them through the wheels of the trucks he lay across the parapet firing until he was hit and knocked back into the trench. He heaved himself and his Bren back into the open again and kept his gun hammering away until he was hit.'

It now seemed as though the platoon must be overwhelmed, but at that moment there was a crash of artillery fire as the guns of 7th Field Regiment RA opened up, driving the Germans back. General Sir Brian Horrocks, then commanding the 3rd Division's machine-gun battalion, 2nd Middlesex, later recalled how the commanding officer of the Royal Ulster Rifles, Lieutenant-Colonel Knox, saw a few of his men, discomforted by shellfire, running back out of the town. After a few words from him, they turned round and began to trot back. '"Wait a minute," he said, "let's have a cigarette." In spite of some fairly heavy shelling he made them finish their smoke. He then said, "now *walk* back to your positions" – and they went.' The defence of Louvain would later gain the Ulster Rifles a DSO for their Colonel, an MC, and four MMs.

By the end of the 15th the infantry holding the Dyle line could feel reasonably pleased with themselves, having repulsed all German efforts to get across the river. But the Belgians to the north of the BEF had been penetrated, and the lightning thrust of von Rundstedt's armour through the Ardennes and across the Meuse threatened the long right flank of the Allied troops who had deployed into Belgium. Withdrawal was therefore the only answer, but, as it had been after Mons in August, 1914, it was difficult to explain the logic of this to those in the front line when they felt that they were getting distinctly the better of the battle. And so began the retreat that would end at Dunkirk.

Sometimes they marched; sometimes they dug fresh positions; sometimes they fought. Soon lack of sleep became much more of a worry than whatever the Germans might have in store. A platoon commander on what it was like to march after 48 hours without sleep: 'We marched without break till ten that morning; then we had a half-hour halt, after which we got a ten-minute halt every hour. This was a doubtful advantage, as ten minutes was too short a time for sleep, too short even to allow one to take off one's boots. Sleep was what we all wanted, and that was what some of the men tried to get. It was all but impossible to try and wake them before we moved off.' Nevertheless, fighting cohesion did not break. Jim Stockman, who fought in the ranks of the 6th Seaforths in the truncated 5th Infantry Division:

'I was proud to say "I am British" watching our efforts at Arras. I have never witnessed bravery like it – and against such staggering odds! A Lance-Corporal with us was limping along, legs shredded, still trying to fire a Bren gun long run out of ammunition. In the end, in sheer desperation, he wrenched the bayonets from two rifles lying on the ground and ran in with four of his men, their bayonets slashing and stabbing in all directions until they were unrecognizable with blood and dirt.'

There were, however, tragedies. With the BEF's southern flank growing longer and more exposed to the threat of the Panzer divisions by the day, Gort, commanding the BEF, was forced to scrape the manpower barrel and commit the ill-equipped and only partially trained 12th and 23rd Divisions. Lacking even anti-tank rifles, they proved no match for the German tanks, but, bewildered as they were, fought until overrun.

By now the decision had been made to evacuate the BEF by sea. Churchill, however, saw it was imperative that Calais be held. To this end 30 Infantry Brigade was ordered from England to defend it. The brigade itself had only been formed in April, 1940, as a scratch formation to reinforce Norway. It consisted of two motor battalions, of which more later, 2nd Battalion 60th Rifles (King's Royal Rifle Corps), and 1st Battalion The Rifle Brigade, together with the 1st Battalion Queen Victoria's Rifles, formerly the 9th Londons* and now affiliated to the

* The London Regiment had lost its distinctive title in 1937, with its remaining infantry battalions being largely renamed as battalions of their parent Regular regiments. The exceptions were Queen Victoria's Rifles, The Rangers and the Queen's Westminsters, although they were retitled as battalions of the KRRC in 1941, the London Scottish (affiliated to the Gordon Highlanders), London Irish Rifles (Royal Ulster Rifles), London Rifle Brigade and Tower Hamlet Rifles, the latter two (each of two battalions) likewise retitled 7th–10th Rifle Brigade in 1941, and the Artists Rifles, also affiliated to the Rifle Brigade, but who

60th. The 1st Rifle Brigade and the 60th were in Suffolk, and the former, after a long day of erecting anti-invasion road blocks, received orders at 1900 hours on 21 May to move to Southampton. They were on the road shortly after 2300 hours and both battalions arrived at the port by midday, Queen Victoria's Rifles having been sent to Dover. The Brigade arrived at Calais on the afternoon of 23 May with orders to operate from there towards the main body of the BEF in order to prevent the latter from being cut off. To help them they were joined by 3rd Royal Tank Regiment, which had recently arrived as part of 1st Armoured Division, now making a belated appearance in France after enduring severe and prolonged equipment shortages. Luftwaffe activity and lack of suitable unloading facilities meant that the battalions were able only to unload part of their equipment and stores.

It soon became apparent that an advance inland was not possible because of German opposition, initially in the form of 1st Panzer Division. Fresh orders were sent from England ordering them to defend Calais, now under German ground and air attack, for as long as possible, but that they would probably be evacuated that night. This was then postponed until the following night.

In the meantime Gort was struggling to anchor the BEF so that the ships could take it off before the Germans cut it off from the coast. Finally, on the 25th, Brigadier Nicholson, commanding 30 Brigade, received orders that he was to defend Calais to the last man in order to divert the Germans from Dunkirk and that there would be no evacuation.

Still the Brigade fought on, forcing the Germans to commit another Panzer division. Eventually, by the afternoon of Sunday, 26 May, having suffered over 50 per cent casualties and totally out of ammunition, the garrison was forced to surrender. Trained for an entirely different type of warfare to the street fighting in which they found themselves engaged, thrust at no notice into a totally unknown situation, subjected to continuous Stuka and artillery attack, which they had no means of countering, as well as attacks by tanks, and finally to be told that they were to be sacrificed, it is remarkable how these three battalions continued to fight on until the end in what was clearly a hopeless situation.

Those who took part were quite clear as to what kept them going. Airey Neave commented: 'It may be fashionable today to sneer at regimental loyalty, but Calais could not have held long without it.' General Tom

---

became no 163 Officer Cadet Training Unit on the outbreak of war, thus fulfilling the same role that they had performed during the first part of the Great War. The old 13th Londons, whose parent regiment was the Middlesex, also retained their title of Princess Louise's Kensington Regiment.

Acton, at the time Adjutant of the 1st Rifle Brigade, when asked why his battalion had fought so well: 'The Regiment had always fought well, and we were with our friends.' There could be no simpler and clearer encapsulation of the regimental spirit.

The evacuation from Dunkirk itself also proved to be a severe test. Try as it might, the RAF could not prevent Goering's Luftwaffe from pounding both the beaches and the ships involved. The troops, too, knew that they had been defeated and that it was only a matter of time before the Germans broke through the perimeter defences. While there were signs of discipline breaking down in some administrative units which had been on the lines of communication, it never did among the infantry battalions. One eyewitness recalled a platoon of the 6th Black Watch marching along the mole singing their regimental song:

> '. . . Of a'the famous regiments that's lyin' for awa',
> Gae bring tae me the tartan o' the Gallant Forty-Twa.'

The 1st and 1st/6th East Surreys found themselves side by side as part of the rearguard desperately holding the perimeter defences and one critical moment saw the two commanding officers operating as a Bren gun team. 'Bala' Bredin of the Royal Ulster Rifles, on his way through the town of Dunkirk to the beaches: 'There were an awful lot of casualties lying around for whom one could do little, many of them from 7 Guards Brigade who appeared to have passed through a short time before. One unfortunate Guardsman, appearing to recognize me as an officer, motioned to me in the light of the flames from the burning buildings. I leant down to see if there was anything I could do. All he said was, "Leave to fall out, sir, please," having said which he appeared quite justified in dying. He had done his duty and maintained the sort of standard to which we all aspire.'

In all Operation DYNAMO succeeded in evacuating no less than 220,000 British and 120,000 French and Belgian troops back to Britain. They expected to be castigated for their defeat by those at home and were surprised at the warm welcome that they received. For those who did not get off the beaches, and these included the seriously wounded, the prospect of prison camp awaited. In general they were reasonably well treated by their captors, but there were two glaring exceptions to this, both involving *Waffen-SS* troops, after their attacks on the Dunkirk perimeter had met with stiff resistance. At Le Paradis on 27 May one hundred surrendered members of the 2nd Royal Norfolks were lined up against the wall of a barn by troops of the SS Totenkopf and mown down by two heavy machine guns. A day later, a part of 2nd Royal Warwicks, 4th Cheshires and some Gunners surrendered to the *SS Leibstandarte* at Wormhoudt. They were herded into a barn and grenades thrown into it, and then the survivors ordered outside in batches of five to be shot. Remarkably, in both cases men survived, at Wormhoudt probably because of the self-sacrifice of a sergeant-major and sergeant who flung themselves on grenades before they exploded. The survivors' testimonies eventually brought the ringleader of the Le Paradis massacre to book and he paid for the atrocity with his life. In spite of war crimes investigations, no one has ever been brought to trial for the Wormhoudt incident, and the suspected perpetrator, former SS General Wilhelm Mohnke, who was to end his war defending the *Fuehrerbunker* in Berlin, lives outside Hamburg to this day.

There was still one final tragedy to be enacted in France. The German invasion had found 51st Highland Division undergoing its tour in the Saarland. Unable to rejoin the BEF it had fought with the French and was eventually driven back to the little port of St-Valéry-en-Caux, which lies between Dieppe and Le Havre, in the hope of being evacuated. Trapped here by Rommel's 7th Panzer Division, and with no ships available to take them off, the Highlanders were eventually forced to surrender. It was a bitter pill to swallow, as Robert Gayler, serving with the divisional machine-gun battalion, the 1st Kensingtons, describes:

'After this things were quiet for some 15 minutes, when suddenly Lieutenant Lavington appeared, looking very haggard. "Destroy your weapons," he said. The four of us behind the guns could not believe our ears. Some of the Jocks got out of the ditch and stood looking at him with their mouths open. He repeated, "Destroy your weapons, the General has ordered us to surrender!" "Not fucking likely, you yellow bastard!" came the reply. "You aren't a bloody rabble," Lavington said coolly, "you're disciplined British soliders and you will obey the orders of your officers. I have personally heard General Fortune [the divisional commander] give the order for immediate surrender." As he spoke a huge tank appeared in the field behind them. It had a black Maltese cross on its side.'

With battalions of all the Highland regiments represented in the Division, no part of the north of Scotland was unaffected by this disaster, although one of its brigades, which had been detached to defend Le Havre, was evacuated.

For a brief period there was serious thought of sending a second expeditionary force to help the French prevent the Germans from overrunning the rest of their country. To this end 3rd Infantry Division was hastily re-equipped, and 52nd Lowland and 1st Canadian Divisions actually began to land at Cherbourg before it became clear that it was too late and they returned to England.

The shattered units returning from France were sent to camps in the Midlands and West Country to reorganize and re-equip. During the next few months 275,000 conscripts were called up, the bulk joining the infantry, since only small arms were in any way abundant. This meant the creation of 120 additional battalions, but their formation could be just as haphazard as that of the Kitchener battalions 25 years before. The future Field Marshal Sir Gerald Templer, returning from a brief post-Dunkirk leave, was told in mid-June that he was to raise the 9th Battalion The Royal Sussex Regiment.

'I have not the slightest idea how it happened, but I was given a very large field on the outskirts of Ross[-on-Wye], with the promise of a few marquees and sufficient bell tents for a thousand men at eight to a tent. That was all. I then went by train to Chichester, the Depot of the Royal Sussex Regiment, dressed as a Royal Irish Fusilier [Templer's Regiment], and reported to the Depot Commander, . . . I said, "Good morning, Sir. I have come to raise the 9th Battalion of your Regiment." He replied, "Don't be such a bloody fool, there is no such thing." I said that was self-evident and this was the object of my visit. He then kicked me out of his office. . . .

Someone told me that the gymnasium was full of regular soldiers of the regular Battalion [2nd] just back from Dunkirk and nobody was doing anything about them. Somewhere I picked up a Sergeant Major by the name of Pack, and we went into the gymnasium and saw them all – I suppose about a hundred – filthy, tired, covered in oil and lying on the floor. I told Pack to fall them in, which he did with a voice like the bull of Bashan, and I told them who I was and that I was raising the new 9th Battalion of their Regiment. I wanted non-commissioned officers, and if they came to me I would nearly kill them with overwork, but as from that moment I would give each NCO one more stripe than he had already got. I also wanted some good private soldiers to fill the rank of Lance Corporal. I then said, "Now anybody who is prepared to take this on, and it is going to be a pretty hard stint, take one pace forward". Every man did so.'

Before July was out the 9th Royal Sussex had their full complement of men, but equipment was still desperately short. In order to provide sufficient rifles and ammunition for his men to have the opportunity to fire early in their training Templer positioned his NCOs close to the local pubs with orders to seize any rifles left in lorries while their passengers had a drink. He also obtained 1,000 rounds of ammunition from a local RAF station in exchange for six bottles of whisky. Formation of an all ranks social club, fostering music, and, not least, lectures in regimental history all helped to bind the Battalion as one.

But while it was important to create a new Army, the more pressing problem was how to deal with the growing threat of invasion from across the Channel. Divisions, most still only partially re-equipped, were deployed to the south and east coasts of England. Beaches were mined, barbed wire entanglements erected, and fire trenches dug. At first all available troops were tied to defending the beaches, but often had excessively long front-ages. Montgomery's 3rd Infantry Division initially had 30 miles from just east of Portsmouth, one of the most vulnerable sectors. Gradually some divisions were pulled back inland to form mobile reserves, and then nine County divisions were formed to man the static defences, giving the Field Army more of an opportunity to begin preparation for the next round with the enemy.

Some men, though, were impatient to strike an early blow at the Germans. Not least was Churchill himself, who in early June harried the Chiefs of Staff with a concept of 'specially trained troops of the hunter class, who can develop a reign of terror first of all on the "butcher and bolt" policy' on the coasts of Occupied Europe. This was the origin of the Commandos, who were formed from volunteers in June and July, 1940. A total of ten were raised, including No 2 Commando, which was specifically formed for airborne operations and would be the forerunner of the Parachute Regiment. Absorbed into the Commandos were a number of Independent Companies, which had been hastily formed in April, 1940, one from each division still in Britain, for service in Norway, the original idea being that they should cover the long exposed coastline in order to prevent the Germans from setting up submarine or air bases. Three Commandos were also raised in the Middle East. This was the beginning of the British 'love affair' with Special Forces and within a couple of years there was a plethora of them, especially in the Middle East. While they undoubtedly performed a valuable service towards ultimate victory, they were often resented by conventional infantry battalions, since Special Forces drew away some of the best of the officers and NCOs. In one case, though, the boot was on the other foot. Regiments and battalions off-loaded some of their 'bad hats' on to No 52 Middle East Commando when it was being formed and they were quick to take advantage of the more relaxed discipline practised by the Commandos. It was only when this was tightened up that matters improved.

In the aftermath of the disastrous campaign in France the Bartholomew Committee had investigated what had gone wrong. One of its major recommendations was for a large number of armoured divisions to be created. This had a major effect on the Infantry. Firstly, 12th and 23rd Divisions, which had suffered so heavily during the campaign, were disbanded, while 42nd Division was converted to an armoured division. In order to man the latter, and help create the other new armoured divisions being formed, no less than thirty-three infantry battalions, none of them Regular, were converted to armoured regiments during 1941–2. They were designated Regiments Royal Armoured Corps (RAC) and were numbered 107–116 and 141–163 Regiments RAC. Many would see action in the various theatres of war, but although they were allowed to retain their regimental badges, worn on a black beret, they resented losing their individual titles. Salt was merely rubbed into the wounds when the Foot Guards were also invited to form the Guards Armoured Division. They were allowed to retain their titles – thus, 1st (Armoured) Battalion Coldstream Guards and 1st (Motor) Battalion Grenadier Guards. All, though, Guards and Line, took to their new role and became as accomplished practitioners of armoured warfare as the Cavalry and Royal Tank Regiment. 1941 also saw attempts to try and obtain a better balance between armour and motorized infantry within the armoured division. To this end a third infantry battalion was added. This had no Bren gun carriers and was known as Lorried infantry. Some armoured divisions transferred the two motor battalions to the two armoured brigades to enable closer co-operation, an organization which became standard in 1942, while others preferred to retain all the infantry in a motor brigade.

While the 12th and 23rd Divisions suffered disbandment, 51st Highland Division, even though it had been lost, apart from one brigade, was resurrected, largely because of the standing it enjoyed in the Highlands on account of its performance on the Western Front during the Great War. Its six Highland battalions which had been forced to surrender at St Valéry, including two Regular (1st Gordons, 2nd Seaforths), were reformed so that it had much the same order of battle as before. The 49th Division, which had fought as separate brigades in Norway, was sent to garrison Iceland, a crucial staging post in the Battle of the Atlantic. While there it changed its traditional flash of the white rose of Yorkshire for the more fitting polar bear on an iceberg. On the other hand, 52nd Lowland Division, which had just missed seeing action in France in June, 1940, began to train as a mountain division in pursuance of Churchill's perpetual desire to re-invade Norway.

Once the threat of invasion had died down in Autumn, 1940, steps were taken by General Sir Bernard Paget, CinC Home Forces, drastically to reform infantry training. Paget wanted to see the infantry become 'the cutting edge of battle' and strove to create an offensive spirit that the

Bartholomew Committee had considered to be lacking in the BEF. Divisional 'battle schools' were set up and these taught what were called 'battle drills', which were designed to make the infantryman react instinctively in the correct way to any given situation. These had originally been developed by General Alexander after Dunkirk. The idea was for troops to carry out tactical training exercises initially at walking pace and then gradually increase the tempo until reaching full tactical speed.

Thus, when first coming under fire the infantryman was taught to apply the acronym 'Down, Crawl, Observe, Sights, Fire', which meant that he should dive to the ground, crawl to cover, observe from where the fire was coming, set his sights and engage with fire. At the section and platoon level, commanders were instructed in the principle of fire and movement, handling their commands as a manoeuvre group and fire group. Thus, on coming under fire, the section commander would, once he had located the enemy and estimated his strength, continue to engage him with fire and look for a covered approach to enable him to attack in flank. Then, leaving his Bren group to provide covering fire, he would take the rest of the section round and assault the position. He was taught to always try and make his assault at right angles to the covering fire, so that the latter could be maintained for as long as possible, and, having taken the position, immediately to adopt an all-round defence, not on the position, since this would probably attract enemy artillery and mortar fire, but forward of it. If the section commander considered the enemy to be too strong for his section, the matter would be referred to the platoon commander, who would adopt the same procedure. As Paget said;

'Battle Drill gives the pattern of how the modern battle should be fought; it gives ideas to subordinate commanders so that they know what to do and how to do it when operating on wide frontages, moving and fighting on their own initiative and yet combining with each other in mutual support, working together to a common plan as do the individuals of a team.'

Company and battalion commanders were taught 'battle procedures', which covered reconnaissance, planning, the giving of orders, fire coordination, and movement, with great stress being laid on what was called 'concurrent activity'. This was designed to save time by ensuring that, while commanders were reconnoitring and planning, their troops would be preparing for the operation.

The divisional battle schools also laid on exercises involving live ammunition in order to get troops acclimatized to being under fire, and a percentage of genuine casualties was accepted in the cause of realism.

Battle drills and 'live fire' exercises undoubtedly helped to instil confidence and overcome the inevitable numbness of mind which coming under fire for the first time causes, but there were occasions, as we shall see, when the battle drills could not be satisfactorily applied. Other steps taken to improve infantry effectiveness were the appointment of a Director of Infantry, who was made largely responsible for doctrine, and large scale manoeuvres, which gave the infantryman experience in working with other arms.

These reforms, however, initially applied only to the troops in Britain. Those engaged with the Axis forces in the Mediterranean and Middle East had to learn through experience.

CHAPTER FIVE

The First Shock of Battle (2)
Middle and Far East 1940–42

WITH Mussolini's declaration of war in June, 1940, the Middle East became an active theatre of war. The main threat came from Libya, where the Italian ground forces numbered some 130,000 men. To face them were the 35,000 men of the Western Desert Force (WDF), whose headquarters was formerly that of 6th Division, which O'Connor had commanded In Palestine as the 7th Division, but which had been renumbered to avoid confusion with 7th Armoured Division. The WDF was built around two divisions, 7th Armoured and 4th Indian. The former was organized as two armoured brigades and a support group, mirroring that of 1st Armoured in England. The support group had two motor battalions and a Royal Horse Artillery regiment. The motor battalions of 1st Armoured Division were, as we have seen, sacrificed at Calais, but those of 7th Armoured Division were also battalions of the 60th Rifles and Rifle Brigade, 1st and 2nd respectively. Indeed, it was appropriate that these regiments should provide immediate infantry support to tanks in mobile warfare, since it accorded well with their traditional skirmishing role. The motor battalion itself was organized as four motor companies, each of which consisted of three motor platoons, with a truck for each, and a scout platoon in Bren gun carriers.

7th Armoured Division itself was deployed to the Egyptian-Libyan frontier immediately on the outbreak of hostilities with Italy, and the motor battalions did their share of patrolling.

'It was a case of bumping over the desert in a truck for some five to fifteen miles, then patrolling for the remaining three to four miles on foot, finding out the information, returning to the truck and driving home.' When the Italians did eventually move in September the Division fell back before them and then watched while the Italians halted sixty miles into Egypt and began to construct a series of fortified camps.

Elsewhere the Italians were more adventurous. Having seized two frontier posts on Sudan's border with Abyssinia on 4 July, a month later they invaded British Somaliland, also from Abyssinia. The garrison of this territory consisted only of the indigenous Somaliland Camel Corps, two battalions of Punjabis, one of the Northern Rhodesia Regiment, and one of the King's African Rifles, with just four 3.7 inch howitzers. There was no way in which they could expect to defend Somaliland successfully in the face of the 25,000 Italians sent against them, even with the hasty reinforcement by the 2nd Black Watch from Aden when the invaders crossed the frontier. Even so, they gave a good account of themselves before being successfully evacuated by sea from Berbera, with Captain Eric Wilson of the East Surreys attached to the Somaliland Camel Corps winning the Victoria Cross for keeping his machine guns in action for four days, in spite of being severely wounded and suffering from malaria, before being captured.

In December, 1940, the Western Desert Force turned the tables on the Italians in Egypt. What began as a raid on their fortified camps evolved into a major offensive, which drove them out of not just Egypt, but Cyrenaica as well. Besides 7th Armoured Division, 4th Indian Division also took part. This had two of its own brigades, each including one British battalion in line with the usual Indian Army practice, and 16 Brigade, with the 1st Queens, 2nd Leicesters and 1st Argylls. The Division also had the 1st Royal Northumberland Fusiliers as its machine-gun battalion. It played a leading role in the first phase, but then O'Connor, commanding the Western Desert Force, was told that it was required in the Sudan and would be replaced by 6th Australian Division, then in Palestine. Thus, after the capture of Sidi Barrani the Indians began to withdraw to the Delta, leaving behind the Northumberland Fusiliers to act as machine-gun battalion to the Australians, and 16 Brigade, now without any supporting arms.

While the Australians reduced the ports of Bardia and Tobruk and advanced along the coast, 7th Armoured Division moved inland, eventually making a celebrated dash to cut off the Italians streaming south from Benghazi. The 2nd Rifle Brigade were in the forefront of what became known as the Battle of Beda Fomm, arriving just in time to cut the coast road, their leading company engaging the Italians as soon as it arrived. The Green Jackets faced a number of uncoordinated attacks, any one of which could have overwhelmed them if it had been pressed, but, as one officer present later commented, 'The Italians were caught on the wrong foot; the Riflemen had their tails up.'

The aftermath was anti-climax as forces were diverted to Greece and their place was taken by fresh untried forces, including the newly formed 2nd Armoured Division, less one of its armoured brigades, which had been sent to Greece. The 7th Armoured Division was also sent back to Egypt to refurbish. It would not be long, though, before active operations resumed in Cyrenaica.

In February, 1941, Erwin Rommel, who had forced 51st Highland Division to surrender in France, arrived in Libya at the head of the battle-hardened *Deutsches Afrika Korps* and the fighting in North Africa took on a very different shape. He soon sent the Western Desert Force reeling back to the Egyptian frontier, with Tobruk, now under siege, left as the only British toehold in Cyrenaica. General Maitland Wilson's W Force, sent to help the Greeks, was also driven back into the sea by another German Blitzkrieg.

The next German target was Crete. The garrison there was commanded by Bernard Freyberg VC, legendary Great War hero, and consisted of Australians and New Zealanders evacuated from Greece, Greeks, Royal Marines, Local Cretan forces, and British troops. Some of the last-named, like the 1st Rangers (9th King's Royal Rifle Corps), were survivors of Greece, but others had been sent from Egypt. These were built around 14 Infantry Brigade, which had been deployed to the island in November 1940, and now consisted of the 2nd Black Watch, veterans of the brief campaign in British Somaliland, 2nd York and Lancaster, and 2nd Leicesters. In addition, there were two other Regular battalions on the island, the 1st Welch, and 2nd Argylls, who had fought at Sidi Barrani. The German attack came as no surprise, Ultra having revealed the details of it, but where Freyberg went wrong was not to accept that the main thrust of the attack would be from the air. Hence, too high a proportion of his force was committed to combating the threat from the sea and not enough to defending the airfields. The fact that the Germans enjoyed total air supremacy did not help. Thus 14 Infantry Brigade had little difficulty in holding its own against the German paratroopers around Heraklion in the north-east of the island, unaware of the loss of the airfields in the west, which enabled the Germans to fly in substantial reinforcements, even after the Royal Navy had successfully intercepted and destroyed a good proportion of the seaborne element.

When 14 Brigade was told that it was being evacuated the news was greeted with disbelief, as, in the words of the York and Lancasters' Regimental History 'to them the whole battle of the last ten days seemed to have been eminently successful'. The Leicesters' officers retired to the cave, which was their mess, and held a farewell dinner before smashing the remains of their wine cellar and crockery and moving out to begin the withdrawal to the docks. Such was the precision with which this was carried out that the Germans had no inkling of it and the complete brigade was at sea well before dawn. With it, however, came the Stukas. Over the next six hours one cruiser and two destroyers were severely damaged with far more casualties to the Brigade than it had suffered throughout the fighting around Heraklion. No wonder a Black Watch piper played a lament from the bridge of one of the damaged ships as she steamed into the harbour at Alexandria that evening. Most of the Argylls, who had been stationed on the south coast, also managed to get away, but the other two British battalions were not so lucky. The 1st Welch were probably the best equipped battalion and Freyberg had made them his force reserve, while the Rangers were in the Suda sector. Both were sent to relieve the now exhausted New Zealanders on the coast near Canea, but were in turn trapped and forced to surrender after a bitter battle, with only some small parties escaping to fight another day.

The only bright spot at this time was in Eritrea. Here 4th and 5th Indian Divisions had invaded from Sudan in January, 1941. By the beginning of February they had closed on the mountain fortress of Keren. There was only one road giving access to it and the Italians had this well blocked. The one answer was therefore a time-consuming operation to seize and secure the features on either side of the road. Among the British battalions involved were the 1st Royal Fusiliers and the 2nd Camerons, after whom Cameron Ridge, one of the first features to be captured, was named. That it took eight weeks to reduce Keren is not surprising. The Official History:

'The weather was hot and the hillsides waterless, and he [the infantry-man] had to clamber up rocks and over huge boulders, plod across shale, and often tear a way through thorn scrub and spear grass. Weapons, ammunition, and the minimum of other necessary equipment made, on a steep slope, an excessive load. During the last hundred feet or so of rocky scramble the supporting artillery had to lift its fire, and an alert defender could man the crest from a covered position behind it and throw showers of small grenades among the exhausted climbers. When, in spite of this, the attacking troops reached the crest, they were often too exhausted and too depleted in numbers to withstand the counter-attacks which were usually made promptly and boldly, supported by the accurate fire of hidden mortars. Moreover, it was a laborious and costly task to supply the foremost troops with rations, ammunition and water, for everything had to be carried up by men, so that perhaps a quarter of a battalion's numbers might be acting as porters at a time when every infantryman was badly needed. Perhaps worst of all was the heart-breaking labour of bringing the wounded off the hillside.'

The capture of Keren was therefore a challenge of endurance, but once it fell Italian resistance in Eritrea ceased.

British Somaliland was regained, also in March, by a force sent from Aden, and the ridding of East Africa of the Italian presence was concentrated on Abyssinia itself. Three Allied forces were now operating. General Alan Cunningham and his South and East Africans were thrusting from Kenya, the Emperor Haile Selassie with his Patriots were advancing towards Addis Ababa from the north-west, while General William Platt's Northern Force, the victors of Keren, pushed from Eritrea in the north.

It was Platt's men who fought the final major battle of the campaign, the seizing of the fortress of Amba Alagi, another mountain fastness, which fell in May, after eighteen days of battle similar to that at Keren. Addis Ababa fell in the same month, but it was November before the final Italian resistance in the country was subdued. Among those who took part in the last phase were the 1st Argylls, veterans of Wavell's first offensive in North Africa and of Crete. Two other success stories were the crushing of a revolt in Iraq in May and the brief campaign against the Vichy French in Syria in June.

Meanwhile, in the North African desert, after two half-hearted British attempts in May and June to breach the Axis lines and relieve Tobruk, the two sides sat and glowered at one another. One activity relieved the monotony for the infantry, and others. These were the 'Jack columns', an idea evolved by the redoubtable Brigadier Jock Campbell, commander of the 7th Armoured Division's artillery. Their object was harassment. They were usually made up of two motor companies, supported by armoured cars and/or tanks, anti-tank guns, some 25-pdr gun/howitzers, Bofors anti-aircraft guns, and Sappers. Philip Gibbs, who fought with 2nd King's Royal Rifle Corps, described how a typical column left the divisional rendezvous

'usually around dusk and making as much ground to the South and West as possible during the first night to get well under and to the rear of the enemy lines. During the day, while in action, the column dispersed and camouflaged up in depressions and wadis. Movement was very limited, due to tracks, footprints etc. Fires were lit during the day under cover and smoke kept down to a minimum . . .

When a target presented itself the column struck hard and fast, causing as much damage as possible, and then melting away in the desert. In fact we had a little motto, which was "Whack Whack and away!"'

These forays were undoubtedly good for morale and helped to maintain an offensive spirit, but this was not enough for Churchill, who wanted more positive and successful action. In July he replaced Wavell with Auchinleck as the theatre commander. Tobruk, however, continued to hold out, although in August, at the request of the Australian Government, its troops began to be relieved, firstly by a Polish brigade and then, in September, by 70th Division. This was formerly 6th Division, which had been reconstituted to command all troops in the Delta and had two British infantry brigades. An army tank brigade also formed part of the new garrison. Life in besieged Tobruk was tough, as the much travelled 2nd Black Watch found:

'The only palliative was that Tobruk had become the symbol and citadel of British doggedness, and for all the discomfort and the muck and the misery was the feeling that it was then the main post of honour open to the British fighting man. Against the one plus, there was nothing but a host of minuses. Water was scarce, and brackish withal; it was drinkable only in tea, and possibly in whisky, if you could get the whisky. Washing was done in sea-water, when it could be got up, which was seldom. Even the NAAFI* was defeated in Tobruk. Once a fortnight they could supply one razor-blade per section of troops and a twist of boiled sweets for every two or three men: cigarettes very occasionally, though a few came up in the ordinary ration. There was a bakery working just outside the town, whose staff did wonders in producing bread almost without a break. Otherwise the diet was bully, tinned fish and biscuits, without intermission, and not much of these. The routine was dull. Patrolling every night . . . for this was pure 1914–18 warfare. Every night there was an air raid, usually over by 10pm; at which hour His Majesty's ships would come sidling in: to nobody is the credit for holding Tobruk more due than the Navy. Day after day there were dust storms, and from these there was no escape; you might huddle in your dugout, or burrow in your concrete pillbox, but the sand would find you out.'

Auchinleck was, however, preparing a major operation to relieve the port, and this, CRUSADER, was launched on 18 November, 1941.

Two weeks of often confused mêlée followed, but Tobruk was finally relieved and Rommel fell back to Gazala and then, by early January, had evacuated Cyrenaica. It was not long, however, before he struck again, catching 1st Armoured Division and then 4th Indian by surprise. Sergeant Grey of the 2nd Camerons, who were still in 4th Indian Division and now

* Navy Army and Air Force Institutes. Their origins lay in the regimental canteens of the 19th century, towards the end of which three enterprising officers set up a Canteen and Mess Co-operative Society designed to centralize the management of these. After the Boer War it fell on hard times and the concept was abandoned until 1916 when the War Office set up the Army Canteen Committee to organize Expeditionary Force Canteens in France and elsewhere. The Admiralty joined the scheme in 1917, and, with the inclusion of the RAF, it became NAAFI in January, 1921, and continues as such to this day.

hardened veterans of the fighting in both Abyssinia and the Desert, gives a graphic description of what it was like:

'About 1530 they arrived. An armoured car and a tank, followed by lorries, came streaming over the top. Twenty Kittyhawks paid no attention and cruised above us, as if on a Bank Holiday. The lorries stopped on the top and the tank and armoured cars came on, watched breathlessly by everyone, until they got neatly picked off by an anti-tank gun as they came round the last corner – nicely within Bren range. The crews only ran a yard or two! Then the party started. The Germans deployed well out of range, got their mortars, machine-guns and a battery going, and rushed over and into our hill without stopping. Meanwhile one could see the infantry dodging about in the bushes on the hillside. Things looked ugly, as presumably we were on a last-man-last-round racket, and it was going to be a night party. No wire and 500 yards to each platoon front.

But as dusk fell word came to thin out at 1915 hours and leave by 1930 hours – a big relief as there seemed to be a lot of Germans. By 1930 hours it was dark and I went out to bring in one of the forward sections. I went off down the hill and saw some people coming my way. So I shouted, "Is that McKay's section?" "Yes!" came the answer. So I went on to tell them to sit on the top of the hill. When I came nearer I felt that something was wrong. Something was! A large German jumped out from behind a bush and pinned me before I could think. Then a German and an Italian officer came running up, took my rifle and equipment off me, stuck automatics in my stomach and back, while the Italian, speaking in perfect English, said, "Lead us to your comrades; tell them to surrender and you will be well treated." I feigned sickness and stupidity and asked for water, but got kicked in the stomach by the German. There seemed no alternative, so I pointed to my left and the German ordered his platoon to go off in that direction, presumably to do a flanking movement. I started off up the hill, with the officers on either side, and stumbling in the dark managed to bring my platoon well on to my flank. Then I aimed for their position, which I could just distinguish in the dark. I heard a Jock say, "Here the b----s come." Then the Italian said, "Shout to them to surrender!" So I shouted "McGeough, McGeough!" (I knew he was a good shot), got within ten yards of them, shouted "Shoot!" and fell flat. The boys shot and got the German in the head and the Italian in the stomach. Grand! So off I ran and rejoined the platoon. By then we were long past our withdrawal time; so back we went, and after a bit of bayonet work by the rear platoon, jumped into lorries and drove off with the Germans lining the road behind us, popping at us at pointblank range.'

There followed a mad scramble back to the Gazala Line, which the remainder of the Eighth Army, as the Allied forces in North Africa were now called, had hastily occupied and begun to fortify.

From the beginning of February, 1942, until the end of May both sides remained static while each prepared to attack. The lay-out of the Gazala Line illustrated the dilemma which faced the infantry in the Desert. The often fast-moving operations were, of course, dominated by the tank. Infantry were often reduced to merely holding ground, but, unless supported by adequate numbers of tanks and anti-tank guns, were very vulnerable. True, the Army tank brigades, equipped with Infantry or 'I' tanks, were dedicated to infantry support, but often these became drawn into the armour battle. The solution at Gazala was to establish brigade 'boxes'. These were wired-in positions, with anti-tank guns and field artillery included in them. The idea was that the armour, which was positioned to the rear, would ride, like the US Cavalry, to the rescue if one should be attacked. There were two snags to this. Firstly, the nature of Cyrenaica meant that there was always an open desert flank, which was there for the turning. In the case of the Gazala Line it stopped at Bir Hacheim, 43 miles inland from the coast and held by the Free French. Even on this frontage, however, the paucity of available troops was such that the boxes were too far apart to be able to give each other mutual support through fire.

These problems were highlighted when Rommel attacked on 27 May. Swinging his armour around the south of Bir Hacheim, he began to engage the British armour. At the same time he made a frontal attack, which isolated the box held by 150 Brigade. This brigade belonged to 50th (Northumbrian) Division, which was the first British infantry division to be sent out from home, and had previously been deployed to Cyprus prior to moving to Iraq in October, 1941, and then joining the Eighth Army in preparation for its projected offensive in Summer, 1942. Up to that time the vast bulk of the infantry in the Eighth Army had been from the Dominions – Australia, India, New Zealand, and South African. Indeed the only other British infantry formations, apart from 70th Division, which was sent to Ceylon and Burma in March 1942, were 201 Guards Motor Brigade (formerly 22 and then 200 Guards Brigade) and 7 Motor Brigade, which had recently been created in 7th Armoured Division by reinforcing the Support Group with an additional battalion, 9th King's Royal Rifle Corps, now refurbished after the Crete disaster.

Rommel succeeded in totally isolating 150 Brigade and establishing a lodgement within the Gazala Line, called the Cauldron by the British, from which counter-attacks failed to budge him. This forced an echelonning back of the Gazala Line so that much of it now faced south to counter the threat presented by the Axis armour. Bir Hacheim, which had gallantly held out for two weeks, was now almost totally isolated, but its Free French garrison was largely rescued, thanks to the efforts of 7 Motor Brigade.

The decisive tank battle at Knightsbridge now took place. Rommel succeeded in trapping the bulk of the British armour. This forced a wholescale withdrawal, which quickly developed into a pellmell retreat as the confusion grew. Something of the atmosphere is reflected in a letter home from an officer in the 2nd West Yorkshires, who were in 5th Indian Division. This had relieved 50th Division in Cyprus, and sailed to Egypt in April. The West Yorkshires had then been sent to Tobruk, having first been given eight 2-pounder anti-tank guns, with which every battalion was now being equipped, but little time in which to train on them. At Tobruk they had to occupy a 9,000-yard section, but 'no sketches or maps showing previous dispositions could be produced for us, although I found out later from people who had taken part in the seige [1941] that these had been elaborately prepared and collated'.

'On 3rd June we were relieved by the [1st] DCLI [Duke of Cornwall's Light Infantry] and on 4th June moved to a spot about 10 miles East of the Cauldron and Knightsbridge. The DCLI had just come from where were are now [Iraq] and had not slept twice in the same place for 30 days. They only stayed in Tobruk one night and were all put in the bag just south of us on 5th June [overrun by part of 15 Pz Div].'

'On the afternoon 4th June we were given orders to do a night advance in MT [Motor Transport] with a dawn attack on an enemy locality. Prior to this our own Artillery Regt had been taken away and loaned to another division. The present CO and I did not take part in the battle, as we were both LOBs [Left Out of Battle] and remained in the B Ech area.'*

'We had to start from our present position . . . at 0324 hrs, and move to the starting line 8 miles ahead in our MT. There we would

---

* The Second World War infantry battalion, as well as armoured, engineer, and artillery regiments, was organized in three parts in the field. F Echelon represented the fighting element, while A Echelon was responsible for keeping F Echelon supplied with the immediate wherewithal to fight the battle. B Echelon, usually commanded by the Quartermaster, contained those elements not required in combat and included those earmarked as 'left out of battle', the cadre on which the battalion could be rebuilt if it suffered heavy casualties.

meet in the dark another Arty Regt whom we had never seen before and who had never seen us, and also a Sqn of Tanks whom we had never met before. All this was supposed to take place in the dark and the attack to start at first light.'

'While we were still embussed, enemy armoured cars attacked the column; nowadays one drives into the attack in lorries and leap out, if you are able to, just short of the objective. Lorries were shot up and burst into flames. . . . Our Carrier platoon drove off the enemy armoured cars, Companies debussed and attacked and captured some of the enemy positions.'

'The Arty Regt which was supposed to have met us on the starting line turned up two hours late. (They had come up with the DCLI and could not find their way about the desert. They were also put in the bag next day.')

'The ground in front of us was supposed to have been cleared of enemy tanks by one of our Armoured Bdes before our attack was put in. This Armd Bde reported the ground clear of enemy tanks, 20 mins later we were being attacked by enemy tanks.'

'The Sqn of Tanks, which was supposed to have remained with us the whole day, departed without even contacting our CO.'

'Our two forward Coys were overrun by enemy tanks and were almost completely written off. Eventually about 20 to 30 ORs came back from each Coy, but only one of the 8 officers, and he was badly wounded.'

The brigade commander ordered the battalion to reform in its B Echelon area.

'Soon after the Bn reached our B Ech area crowds of vehicles, armoured cars and guns came pouring back through us and to our flanks. [An] officer of the DCLI dashed up in a truck and said that the rest of his Bn was in the bag and the Hun was coming. The CO then sent me to Bde to find out what was happening. Bde said that there was nothing to worry about, so I returned and informed the CO. By this time the flap was reaching prodigious dimensions, with people going almost too fast to stop and tell one what was happening. So the CO sent me off again to Bde. They, though very apprehensive, said there was nothing to worry about and Div confirmed this. As I drove back to the Bn I saw more and more vehicles pouring back. The whole desert was alive with them, kicking up a lot of dust. This and the setting sun made it hard to tell what was what.'

'As I reached the Bn area shells were landing round about and Jerry tanks were in sight. I found no Bn except for six of our carriers and two At [anti-tank] guns (the remainder having been knocked out in the morning) and these were just moving off.'

The surviving members of the Battalion, 'who had lost everything except what they stood up in', were eventually reunited next day at the airfield of El Adem, eight miles east of Tobruk. They were then sent back to the Nile Delta to refit.

The retreat now degenerated into a virtual rout. Tobruk, held by the South Africans and which Auchinleck hoped would hold, quickly fell and the Eighth Army passed back across the frontier, through Mersa Matruh to a 'last ditch' defensive line at El Alamein. This had been under preparation for some time, but the 2nd West Yorkshires, who were sent to occupy a sector in the extreme south, on the edge of the Qattara Depression, found 'nothing except for an empty water cistern and a few holes and slit trenches, which were full of excreta'. Meanwhile the Mediterranean Fleet left its base at Alexandria for Haifa, and Headquarters Middle East in Cairo began to burn its classified papers and prepare for evacuation across the Suez Canal to Palestine.

If the British situation in the Middle East was grim during the first six months of 1942 so it was elsewhere. The Japanese bombing of Pearl Harbor in December, 1941, may have fulfilled Churchill's earnest desire to see the United States enter the war, but it resulted in an oriental blitzkrieg which drove the Americans, British and Dutch out of almost all their territories in the Far East.

The Japanese struck at Hong Kong and Malaya on 8 December. The garrison of the former was built around six infantry battalions, two Canadian, which had only recently arrived, two Indian and two British. The last-named were the 2nd Royal Scots, who had spent many years overseas and had been due to return home in 1940, and 1st Middlesex, who were a machine-gun battalion. Three Japanese divisions took part in the invasion and the only hope for the defenders was that Chinese troops might arrive to relieve the pressure, but this was little more than wishful thinking. The New Territories were totally in Japanese hands within five days and they then began to attack Hong Kong Island systematically from the air and with artillery. They began to land on the island on the night 18/19 December and, in spite of often fierce resistance, the garrison was forced to surrender in the early hours of Boxing Day. By then the Royal Scots were down to a mere four officers and ninety-eight men,and the Middlesex were in a little better state. It is a measure, however, of how both battalions conducted themselves under the harsh conditions of being prisoners of the Japanese that both Lieutenant-Colonel 'Monkey' Stewart of the Middlesex and Captain Douglas Ford of the Royal Scots were later awarded the posthumous George Cross for their fortitude.

The fall of Malaya and Singapore was an even greater disaster. On paper the garrison looked strong, with two Indian and one Australian divisions, each of two brigades, two independent Indian infantry brigades holding Malaya, and a further two infantry brigades making up Singapore Fortress. The cream of the Australian and Indian troops were, however, in the Middle East, and it was for desert warfare that many of the troops in Malaya had originally been trained. Hurriedly sent from India they had little time to become acclimatized. One of the few British battalions present* was the 2nd Argylls, who had been sent to Malaya in 1939 after a tour in India, which included two frontier campaigns. They were one of the few to take jungle training seriously, in the face of local opinion which considered most of the jungle 'impassable to infantry'. Captain Angus Rose of the Argylls:

'The casual observer got the impression that the whole jungle was one dense mass of impenetrable foliage . . . this was not the case. Along the borders of the roads, where the trees had been cut down and light had come in, the jungle grew up again with great density. This was known as secondary jungle. Once inside this was a narrow belt, which went only to a depth of a score of yards, there was the real, or primary, jungle. It was not normally necessary to cut through primary jungle, except in river courses or over water-logged ground . . . Tourists and office-bound staff officers, who never had time to leave the road, got a completely wrong impression of what the jungle really was and "impenetrable jungle" was, of course, a complete myth.'

Defence was therefore mainly based on the roads, and too often the Japanese merely took to the jungle and outflanked positions. There was also a dangerous under-estimation of Japanese capabilities. As one British soldier commented: 'We never considered the Japs were people worth bothering about.' One symptom of this was the total lack of training in anti-tank tactics until after the Japanese had attacked, and by then it was too late. This was in the mistaken belief that the Japanese were unlikely to employ tanks.

The Japanese invasion began with three simultaneous landings on the east coast of Thailand and Malaya. In the space of some seven weeks the Japanese totally out-manoeuvred the defenders, driving them down the

* The others were 2nd Loyals, 1st Manchesters and 2nd Gordons in Singapore Fortress, and 2nd East Surreys and 1st Leicesters in 11th Indian Division. Contrary to normal practice, 9th Indian Division had no British battalions.

400 miles of the Malayan peninsula and across the Causeway to Singapore Island. While there were a number of gallant actions, the constant withdrawals had their effect. An eyewitness:

'The troops were very tired. Constant enemy air attacks prevented them from obtaining any sleep by day. By night they either had to move, obtaining such sleep as was possible in crowded lorries, or had to work on preparing yet another defensive position. The resultant physical strain of day and night fighting, of nightly moves or work, and the consequent lack of sleep was cumulative and finally reached the limit of endurance. Officers and men moved like automata and often could not grasp the simplest order.'

Because of casualties, the 1st Leicesters and 2nd East Surreys soon had to be amalgamated as the 'British Battalion' in 11th Indian Division, and the Argylls, on withdrawal to Singapore, absorbed a number of sailors, survivors of the *Prince of Wales* and the *Repulse*, and became known as the 'Plymouth Highlanders'.

Just as the evacuation of Malaya for Singapore was beginning, reinforcements in the shape of Indian, Australian and British troops began to arrive. The last comprised 18th Infantry Division, which was made up of Territorial battalions of the Royal Norfolks, Suffolks, Sherwood Foresters, Bedfordshire and Hertfordshire Regiment, and the all-TA Cambridgeshires. They had originally left England at the end of October, 1941, sailed across the Atlantic to Halifax, where they had been transferred to American transports, even though the USA was still not officially at war, and sailed to Cape Town. They were earmarked for the Middle East, but news of the crisis in Malaya caused a change of plan. Most of the division went on to Bombay, but one brigade, after stopping off at Mombasa for a week, went direct to Singapore, arriving there on 13 January. Jim Bradley, a member of 53 Infantry Brigade, recalls that on disembarkation:

'We were given an official printed document, telling us how to differentiate between Japanese, Chinese and Malays. The one significant difference, we were told, was that the Japanese always walked with their toes turned in, but this was very little help when they were driving a truck or tank, or even riding a bicycle! . . .'
'This rather useless document also gave us particulars of how payment would be made to the families and dependents of prisoners of war, should Singapore fall; not a particularly encouraging start to our service overseas!'

The remainder of the division arrived a few days later, but, with no chance of acclimatize after some three months in transit, there was little that it could contribute. On 8 February, 1942, the Japanese landed on Singapore and a week later it was all over.

A few intrepid spirits managed to escape from Singapore, and, in one case, an officer was ordered off the island just before it surrendered. This was Lieutenant-Colonel Ian Stewart, who had commanded the 2nd Argylls and temporarily 12 Indian Infantry Brigade. His jungle warfare expertise and experience were considered so valuable that he was, unusually, not allowed to go into captivity with his men. The fate of the vast majority was to be condemned to work on the infamous Burma-Siam railway. Something of what they experienced is given in this diary entry by Robert Hardie, who was a doctor on the Railway. The time is May, 1943:

'The accommodation for the men in this camp is hopelessly inad-equate; the tents are crammed, but still men have to sleep in the open (some prefer to) or under such primitive shelters as they can improvise with bamboo and a little atap and perhaps a groundsheet. They are being worked hard too. They parade after a hasty breakfast about three-quarters of an hour after dawn, and go to 6 or 7 Tokyo time (to within an hour or two of sunset) bamboo cutting, tree felling, bridge building, embankment building and making cuttings, pile driving and so on, all in blazing sun under constant pressure backed up by violence ... The sickness in 16 Battalion [largely made up of 5th Royal Norfolks] in these six weeks has become alarming – 240 out of 400 unable to work now. Many are desperately ill with dysentery, beriberi and pellagra, malaria and exhaustion.'

Many did not survive and those who did were undoubtedly helped by the fact that the Japanese, while they often separated the officers from their men, did not generally break up battalions and their structure survived.

The other major disaster which the British suffered at the hands of the Japanese was the loss of Burma. The story here was much the same as in Malaya, but the British had fewer troops to hold a very much larger area. Thus the Japanese infiltration tactics through the jungle were even more successful. Again, there were few British battalions involved, with the defence of Burma being initially entrusted to 1st Burma Division, which included 1st Glosters and 2nd King's Own Yorkshire Light Infantry, both of whom had been depleted by having had to send a number of their officers and men back to Britain. The Burma Division was reinforced in early January by 17th Indian Division, which included two further British Battalions, 1st Cameronians and 1st Royal Inniskilling Fusiliers. After the fall of Rangoon in early March all were involved in the long retreat northwards and back into Assam. All four became severely depleted in men, as they struggled to clear the many blocks that the Japanese placed in their path. Exhaustion, too, took its toll. As one Inniskilling officer said to an American journalist after an attack on a strong Japanese position in a village: 'Our men were worn out, but they had guts. They crouched low with bayonets fixed and charged forward like the Guards on the parade ground – or, at least, tried to. The plain fact is that they were so worn out that they stumbled forward like drunken men hardly able to hold their rifles. But they gave them hell . . . and drove the bastards out of Twingon.' Yet, they were able to prevent the Japanese from achieving their aim of destroying the Burma Army before it extricated itself from the country. General Bill Slim, who had commanded Burcorps, comprising 1st Burma and 17th Indian Divisions and 7th Armoured Brigade, whose North African veterans acted so ably as the rearguard for much of the retreat until forced to leave their tanks on the other side of the Chindwin, wrote:

'On the last day of that nine-hundred-mile retreat I stood on a bank beside the road and watched the rearguard march into India. All of them, British, Indian and Gurkha, were gaunt and ragged as scarecrows. yet, as they trudged behind their surviving officers in groups pitifully small, they still carried their arms and kept their ranks, they were still recognizable as fighting men. They might look like scarecrows, but they looked like soldiers too.'

Unlike those who arrived back from Dunkirk in Summer, 1940, there was no warm welcome. Amid the teeming monsoon rains the survivors were merely directed to swampy bivouac areas, which lacked medical arrangements, water supplies, even tents and blankets. The result was, in Slim's words, 'I should estimate that eighty per cent of the fighting men who came out of Burma fell sick, and many died.'

Thus, at the end of June, 1942, in both theatres of war in which the British Army was actively engaged it had suffered disastrous reverses. The way ahead appeared long and very uphill.

CHAPTER SIX

Turning the Tide
Mediterranean Theatre 1942–43

By 1942 the growing complexity of land warfare meant an increasing number of specialists in the British Army. It was thus important that new recruits were fitted as round pegs into round holes. For this reason the General Service Corps was established to screen all new conscripts. These now all went through eight weeks of basic training at what were called Primary Training Centres (PTC). These were attached to Infantry Train-ing Centres (ITC), which had taken over from the Regimental depots, soon found to be too small to cope with the numbers of recruits. While at the PTC the conscripts underwent various tests and then were allotted to the arm or service considered the most suitable for them. Those earmarked for the infantry remained at the ITC for a further eight weeks of training and were then posted to a home-based battalion. From there they would either go overseas with it, if it was designated a Field Force battalion, or be sent as part of a reinforcement draft to a battalion already abroad. The battalions restricted to service in Britain were designated by various names. There were Home Defence battalions *per se*, Reserve battalions (later called Holding battalions), and Young Soldiers' battalions. Within Scottish Command, however, two new regiments were formed in 1942, the High-land and the Lowland Regiments, to either of which each Scottish infantryman was badged while he did his special-to-arm infantry training. These were the only two new infantry regiments formed during the Second World War and would not be disbanded until 1949.

There were, too, a number of measures taken at this time to improve the combat effectiveness of the infantry battalions. One of the most significant began to be introduced towards the end of 1941. This was radio, termed 'wireless' in those days, with a range of sets coming into service. The No 38 worked from platoons to companies, the No 18 from company to battalion, and the No 22 from battalion to brigade. These were all high frequency sets, and battalions operating in South-East Asia did not use the No 38, since it seldom worked in the jungle; the Nos 18 and 22 were often unreliable for the same reason.

The Thompson sub-machine gun gave way to the Sten, cheap to produce and lighter than the 'Tommy Gun'. Likewise, during 1941 the SMLE rifle began to be superseded by the No 4, which had the same 0.303 inch calibre and a similar mechanism, but was, like the Sten, much cheaper to produce. With it came a new bayonet, only half the length of the traditional 18-inch version. Veterans were doubtful about both at first, but soon grew used to them, and, while the new bayonet certainly appeared much less formidable, it did mean, with its significantly reduced weight, that the rifle was much easier to handle and more accurate to fire when the bayonet was fixed.

By early 1943 the Boys anti-tank rifle was being phased out, with little regret, and replaced by the Projector Infantry Anti-Tank, more commonly known as a PIAT. This fired a hollow charge bomb, and its effectiveness was clearly demonstrated by Fusilier Francis Jefferson of the 2nd Lancashire Fusiliers in Italy in May, 1944. During a German counter-attack and under heavy fire he leapt out of his trench and engaged two Tiger tanks, firing his PIAT from the hip and knocking out one and forcing the other to withdraw. For this he was awarded a well-earned Victoria Cross. Even so, the PIAT was not wholly satisfactory; it was cumbersome to carry, difficult to cock on account of the strong spring which propelled the bomb, and had a violent recoil.

The battalion's mortar platoon was increased to six mortars, and was mounted in Bren gun carriers. Likewise, an anti-tank platoon was established, initially with 2-pdr anti-tank guns, but later with six 6-pdrs. The pioneer platoon was also converted into assault pioneers, who were trained in demolition work. These additions to the Headquarter Company made it unwieldy, and hence the support weapons were concentrated in a new Support Company. A final amendment was the establishment of two snipers per company, later concentrated in a sniper section at Battalion HQ. British battalions serving in India remained on the pre-war establishment, however, and their only support weapons were six 3-inch mortars.

Turning to the progress of the war, British attention in Summer, 1942, was largely fixed on North Africa and Eighth Army's attempt to prevent Rommel from reaching the Suez Canal. The struggle at El Alamein in July, 1942, was like that of two exhausted boxers in the last round of a marathon contest. The beginning of it saw Rommel striking groggy blows at his opponent, only to have them parried. His one success was in overrunning 18 Indian Infantry Brigade, newly arrived from Iraq, with a temporary commander and some hastily gathered anti-tank guns and 25-pdrs, on 1 July, the first day of the battle. Even so, this brigade, which included the 1st/5th Essex, held him up until the evening. It was now Auchinleck's turn, and throughout most of the remainder of the month he rained blows at various points in the Axis line, but none was able to achieve a breakthrough.

The main problems were shoddy staff work, poor communications, and a lack of co-operation between the tanks and the infantry. These, however, were indicators of an army that was exhausted after two months of non-stop fighting. Even so, Rommel had been stopped and the Eighth Army now had the chance to pause for breath and reorganize.

The replacement of a tired Auchinleck by the new team of Alexander and Montgomery in mid-August infused the Eighth Army with a fresh spirit. Monty made it clear that whatever happened there would be no further withdrawals and that the Eighth Army was to begin to prepare for an offensive. Before this Rommel struck once more at the El Alamein line, but Monty, warned by Ultra, was ready for him. The main clash came just to the west of the Alam Halfa ridge, and involved 22 Armoured Brigade, which had the 1st Rifle Brigade as its motor battalion, and two brigades of the recently arrived 44th (Home Counties) Division. The tanks and the Rifle Brigade's 6-pdr anti-tank guns gave Rommel's armour a hot reception and repulsed it. Matters, however, did not go so well on the night of 3/4 September, when 132 Infantry Brigade of the Home Counties Division and 5 New Zealand Brigade launched an attack to try and close the minefield gaps to the rear of the *Deutsches Afrika Korps* and thus trap it. The 2nd Buffs and 4th and 5th Royal West Kents were an hour late crossing the start line, by which time the Germans were thoroughly alerted and suffered heavy casualties in what was their first action. They had 697 men killed, wounded and missing, including their brigade commander severely wounded. After this, though, the Germans withdrew behind the British minefields and Montgomery began to prepare for his major offensive designed to 'knock Rommel for six right out of Africa'.

The artillery barrage that preceded the Eighth Army's attack at El Alamein on the night of 23/24 October was the largest fired by the British Army since 1918, and was of much reassurance to the infantry of the four infantry divisions which were to carry out the break-in operation through the minefields. Three of these divisions were from the Dominions, and veterans of the desert war – 9th Australian, 2nd New Zealand, and 1st South African. The other was the reconstituted 51st Highland Division, whose opportunity to revenge St Valéry had finally arrived. One of the subalterns of the 7th Argylls, H. P. Samwell, whose first battle this was, recalled the day of the 23rd, which his battalion spent lying uncomfortably in temporary slit trenches just behind the front lines:

'The tension was almost unbearable and the day dragged terribly. I spent the time going over and over again the plan of attack, memorizing codes and studying the over-printed maps which showed all the enemy positions – the result of weeks of patrolling and reconnaissance from the air . . . Oddly enough, although keyed up, I did not feel any fear at this time, rather a feeling of being completely impersonal, as if I were waiting as a spectator for a great event in which I was not going to take any active part.'

Once darkness had fallen a hot meal was brought up for the attacking troops, accompanied by last-minute minor adjustments to the plan. At 9pm they got out of their trenches, moved forward, and then lay down as the barrage opened. Twenty minutes later the attackers began to move forward, preceded by Sappers with mine detectors to clear lanes through the minefields. Samwell recalled the enemy machine guns and mortars opening up as they closed, and soon:

'The line had broken up into blobs of men struggling together . . . I saw some men in a trench ahead of me. They were standing up with their hands above their heads screaming something that sounded like "Mardray". I remember thinking how dirty and ill-fitting their uniforms were, and smiled at myself for bothering about that at this time. To my left and behind me some of the NCOs were rounding up prisoners and kicking them into some sort of formation.'

By dawn the 7th Argylls had secured their objective and had dug in, ready for any counter-attack. Elsewhere, though, the initial objectives still remained to be taken and Montgomery's timetable began to slip. The armour's attempts to break through resulted in heavy casualties, and on 27 October the Axis forces began to counter-attack. One battalion which found itself facing the Axis armour was the 2nd Rifle Brigade, positioned on a small feature called 'Snipe', south-west of the vital Kidney Ridge. Their 6-pdr anti-tank guns, which they had only recently received, supported by those of 239 Anti-Tank Battery, repulsed numerous attacks, beginning in the early hours of the morning and continuing into the late afternoon. Sergeant Charles Calistan MM was in charge of one of the Rifle Brigade's guns:

'We were giving them hell, but we weren't by any means getting away with it. Our position was rather exposed and they let us have everything they had got. They even attacked us with lorried infantry . . .'
'All this time the enemy never let up; nor did we. Time seemed to be lost in the battle. My gun smashed up five tanks in that first attack – and I am only counting those that "brewed up", that's our way of saying they burnt out. Some of our guns were out of action. Some had run out of ammo . . . The thing that sticks out is the Company

Commander saying that we were cut off and there wasn't anything that could get through to us. We would fight it out. Keep on firing as long as we had shell or bullet! Yes. We understood.'

'And when you had time to listen, it was only then when you realized that you had fewer and fewer guns firing. We were also short of water, but somehow you didn't think of that. Two of my gun crew crept out on their bellies – right into the open to get some ammo. They were under enemy fire the whole time and their progress was terribly slow. Then our platoon officer decided to reach his Jeep, which had four boxes of ammo on board. God knows how he got to it – they were machine gunning the whole way. He started coming towards us and then they hit the jeep and it caught fire, but he kept on coming. We got the ammo off, and then I had an idea. We hadn't had a thing to drink and we naturally hadn't been able to light a fire, but here was a perfectly good one. So I put a can of water on the Jeep and it brewed up well enough for three cups of tea!'

'Our Colonel [Lt Col V. B. Turner] kept going from gun to gun. How he inspired us! The enemy tried to shift us with an infantry attack, but we soon sent them on their way with our Bren-carriers and our infantry, who were in position in front of us. When the next tank attack came in, the Colonel was acting as loader of my gun. He got wounded in the head – a nasty wound and we wanted to bind it up, but he wouldn't hear of it. "Keep firing!" – that's what he wanted, and we didn't pause. When the gun ran short of ammo he got it from one of the others.'

That evening the Battalion was withdrawn, unfortunately forced to leave its guns behind, although the surviving crews took the breech-blocks with them. They and their supporting Gunners had suffered seventy-two killed and wounded, but subsequent careful investigations revealed that they had accounted for no less than fifty-seven enemy tanks, guns, and other vehicles. Their resolute defence demonstrated that infantry equipped with effective anti-tank weapons could hold off armour, just as their forbears had successfully resisted horsed cavalry with their squares. Their performance resulted in the award of the Victoria Cross to Vic Turner and many other decorations for bravery. These included a Distinguished Conduct Medal to Sergeant Calistan, who was also recommended for the VC. Subsequently, he was commissioned but was killed in action in Italy.

Lack of decisive progress forced Montgomery to change his plan, and in the early hours of 2 November he struck at a new point in the defences, just north of Kidney Ridge. Spearheading this new attack were the 8th and 9th Durham Light Infantry of the now battle-hardened 50th Division and

1. Men of the Wiltshires celebrate a successful trench raid, Spring, 1916. Note the rum jar in the foreground.

2. Men of the 1st Black Watch just before the great German offensive of March, 1918. The goatskin jackets worn by some men were popular during the first winter on the Western Front, but were then largely replaced by leather jerkins.

3. 6th York and Lancs at Gallipoli, Summer, 1915. The two men awake
are trying to tempt a Turkish sniper into firing so that he will reveal
his position.

4. A subaltern leads a wiring party of the 12th East Yorkshires up a communication trench, early 1918.

5. 1st/9th Hampshires parade in Vladivostok prior to entraining for Omsk to support the White Russian forces, December, 1918.

6. British battalion on a route march on the North-West Frontier of India, mid-1930s. Frontier soldiering was excellent for honing basic infantry skills.

7. The Phoney War. General Georges, commanding the Allied North-Eastern Group of Armies, inspects a Guard of Honour provided by the 2nd Royal Norfolks, France, Autumn, 1939.

the 5th Seaforths and 5th Camerons from 51st Highland. Under a barrage even smore intense than that which had opened the battle, they moved forward amid 'the crump of mortars . . . and, ripping through everything, the crack of Bredas and the vicious pup-turrr, pup-turrrr of the Spandau, the German light machine gun'. The experience of the 5th Seaforths:

'Clouds of dust and smoke arose, blotting out the desert which so short a time ago, had seemed so vast. Each man found himself in a diminished world inhabited by himself and at most three or four others. Somewhere near him was an officer or sergeant with a compass, trying to walk a straight course through the inferno, for more than two miles, to an objective that was only a pencil line on the map. Inevitably, groups divided and subdivided. Some companies remained intact, moving on their appointed courses. Others lost a Section or two. Some lost a Platoon. But there was a compass in nearly every group, and most of those who survived the barrage did reach the objective. It was, unfortunately, a question of surviving the barrage. Through inexperience, the forward Companies took too literally the injunction which had been drilled into them, and kept so close that, until they reached the half-way line, they were under our barrage instead of behind it. German shells and mortar bombs were landing too, and casualties were heavy.'

The cost to the Battalion was twelve officers and 165 other ranks killed or wounded, but by dawn they had, like the other battalions, reached their objectives, and the tanks of 1st Armoured Division pushed through. That evening Rommel, having launched a failed counter-attack against the breach, realized that a breakthrough was inevitable and began to withdraw. Perseverance, especially by the infantry, had finally won through and the stage was now set to drive Rommel out of Egypt and Libya.

As the Eighth Army began its pursuit of Rommel, 2,000 miles to the east the first major amphibious landings to be carried out by the Western Allies were taking place. These were the TORCH landings on the French North African coast. Apart from the numerous Commando raids, including the disastrous August, 1942, Dieppe operation, which had also involved a Canadian infantry division, there had only been one amphibious landing which had been more than just a 'tip and run' attack. This was the invasion of Vichy French Madagascar, which had taken place on 5 May, 1942. The nucleus of the force was 29 Infantry Brigade, which had been formed in early 1941 as part of Force 110, which was created as an amphibious expeditionary force. As such, 29 Brigade had received much training in amphibious warfare, but it was joined shortly before the expedition set sail from Britain by two brigades of 5th Division, 13 and 17 Brigades. All three were designated infantry brigade groups, indicating that they had their own supporting arms, consisting of a field regiment RA and field company RE. These two brigades had been earmarked, with their parent division, for India, and consequently had no opportunity to develop the necessary specialist skills before sailing. Indeed, it was not until the force reached Durban that 17 Brigade, which was earmarked for the second wave of the landings, was able to carry out any proper training. The Commanding Officer of the 6th Seaforths in this brigade*, Lieutenant-Colonel G. S. Rawstone was a firm believer in physical fitness. According to Jim Stockman: 'He ordered that the battalion should form every day and march from bottom to top of the ship and completely round it – and this with full pack and in sticky tropical heat. We swore and we moaned at the time.' The landings, carried out by night on the north of the island, took the French by surprise. Even so, once ashore and advancing inland, the leading troops of 29 Brigade, 1st Royal Scots Fusiliers and 2nd Royal Welch Fusiliers, did meet opposition. Nevertheless, they closed to the port of Antsirane by nightfall. It was now that the physical fitness training of the 6th Seaforths paid off. They and the 2nd Northamptons had to make an 18-mile forced march to reinforce 29 Brigade and then pass through them and carry out an assault on the town. This was carried literally at the point of the bayonet and northern Madagascar was soon pacified. Shortly after this, however, that scourge of tropical climes, malaria, struck without warning, the 6th Seaforths alone suffering nearly 250 cases. The force was therefore relieved by better acclimatized East African troops, who completed the securing of the remainder of the island.

The TORCH landings themselves took place on 8 November, but the bulk of the initial assault was carried out by US troops, since it was rightly considered that the French forces would be more likely to down their arms willingly to them than the British, whom they had not forgiven for the bombardment of the French fleet at Mers-el-Kebir and Oran in July, 1940. Once the French in Morocco and Algeria had joined the Allies, eyes turned on Tunisia, and it was now that the British began to take a leading part. The only troops involved in the initial landings had been Nos 1 and 6 Commandos, but within a very few days they were joined by a Parachute brigade, with the first three operational Parachute battalions in the British Army, and 78th Division. This was a new formation, created earlier in the year, and initially consisted of just two brigades, 11th and 36th. To them fell the task of spearheading the Allied advance into Tunisia.

* The other two battalions were 2nd Royal Scots Fusiliers and 2nd Northamptons. 29 Inf Bde Gp consisted of 1st Royal Scots Fusiliers, 2nd Royal Welch Fusiliers, 2nd East Lancashires, and No 5 Commando.

Accompanied by small Commando operations on the coast and airborne landings on various airfields, 11 and 36 Brigade set forth into Tunisia, crossing the border on 15 November. In the meantime the Germans had reacted and began to fly reinforcements into Tunisia. The result was an encounter battle. The first clash was between a German battle group commanded by Major Rudolf Witzig, who had led the spectacular German glider assault on the Belgian fortress of Eben Emael on 10 May, 1940, and Hart Force, the advanced guard of 11 Brigade, and which consisted of a company and carrier platoon of the 5th Northamptons, and a squadron of 56th Reconnaissance Regiment.*

Hart Force soon realized that Witzig's force was too strong and reinforcements were summoned in the form of the 6th Royal West Kents, who had been landed by destroyer at Bône. This battalion had a distinct sporting character. Its commanding officer, Lieutenant-Colonel 'Swiftly' Howlett, had played cricket for Kent as a fast bowler, and his anti-tank officer, B. H. Valentine, had been his County captain. They had only four 2-pdr anti-tank guns and some Boys anti-tank rifles with which to face some 17 Pz Kw IIIs and IVs, but with help from their supporting 25-pdrs they succeeded in holding them and the German infantry off for most of a day.

Better progress was made by 11 Brigade and on 27 November the 1st East Surreys reached Tebourba, just 20 miles from Tunis. They, too, had to repulse a tank attack, and the following day the 5th Northamptons, supported by American tanks, passed through them to tackle the next objective, the village of Djedeida, which lay just 15 miles west of Tunis. They made two attacks on the German positions there, but were beaten back, largely thanks to the fact that the Germans enjoyed total air supremacy at this time. The Northamptons withdrew to a position 2,000 yards east of Tebourba and were relieved by the 2nd Hampshires.

The Hampshires had landed at Algiers only on 22 November and were part of the third brigade to arrive in Tunisia, 1 Guards, which also consisted of the 3rd Grenadiers and 2nd Coldstream. The Hampshires were, however, now detached to under command 11 Brigade. The defen-

* The Reconnaissance Corps had been formed in January, 1941. The object was to replace the Cavalry regiments which provided the reconnaissance element of the infantry divisions, since the former were required for the new armoured divisions then being formed. They were, in essence, armoured car regiments, and had a similar organization. They were initially formed from existing infantry battalions, but on 1 January, 1944, became part of the Royal Armoured Corps, taking on its mailed fist and surrounding winged thunderbolts capbadge. The Recce regiments' numerical titles reflected the infantry divisions of which they were part. Thus, 56th Recce Regt was initially part of 56th (London) Division, but changed its title to 78th Reconnaissance Regiment, while in Tunisia. The Reconnaissance Corps was disbanded in August, 1946. For the first comprehensive history of the Recce Corps see Richard Doherty *Only the Enemy in Front* (Tom Donovan, London, 1994).

sive position hastily taken up by the Northamptons was not very satisfactory, and Lieutenant-Colonel James Lee, commanding the Hampshires, wanted to adjust it, but was told that he could not. In fact, 11 Brigade, poised for another thrust on Tunis once additional Allied troops arrived, was poorly placed for defence, with the Hampshires two miles ahead of the East Surreys, who, in their turn, were four miles ahead of the Northamptons. Matters had been made worse by the incessant German air attacks. Consequently, on 1 December, when the Germans began to counter-attack from the south-west the situation soon became serious. It first hit Blade Force, a battle group built around the Crusader tanks of the 17th/21st Lancers. These only had 2-pdr guns, significantly inferior to the those of the German tanks, and were soon brushed aside. The Germans then turned on the two forward Hampshire companies. Early on the 2nd, supported by machine guns which overlooked its position, they succeeded in overrunning one of the forward companies, but the remainder of the battalion held throughout the day. That night the Hampshires were given orders to withdraw to the Tebourba area, which they did.

On 3 December the pressure switched to the East Surreys. A furious attack knocked a company off the vital Pt 186, and a gallantly mounted counter-attack by two other companies wilted under heavy fire. Their position now untenable, the East Surreys were also ordered to withdraw. This was done at night, amid heavy fire; to one officer 'it was Dunkirk all over again', but 'fortunately the Boche were using tracer, and it was not difficult to avoid'. This left the Hampshires on their own at Tebourba, with both flanks now entirely exposed. Major Herbert Le Patourel led a series of desperate counter-attacks to try and restore the situation, the last being just himself and from which he did not return. He was later awarded a posthumous Victoria Cross, but did, in fact, survive, although badly wounded, and endured the rest of the war as a POW. The remainder of the battalion, now down to 200 men, formed a square around battalion headquarters and made a series of bayonet charges to try and relieve the pressure. Then, after dark, with ammunition and water low, their anti-tank guns and supporting 25-pdr battery destroyed, Colonel Lee called the survivors together and told them that they would have to break out. He ordered them to fix bayonets and to 'walk forwards and when you get close enough charge and give it 'em'. This they did, led by Lee and his adjutant, both already wounded. Reaching Tebourba, now empty, the Hampshires marched through in column of threes, their colonel at their head. Beyond it, it became clear that the road to Medjez el Bab was cut. Lee therefore told his men that they could surrender if they wanted to, but none took this up. He therefore ordered them to form groups of two or three and make their way back to Medjez through the hills.

By 6 December four officers and 120 men had reached the rendezvous east of Medjez. Lee was not among them; like Le Patourel, he had been captured, but was later awarded the DSO for his superlative leadership. After interviewing the survivors, Evelyn Montague of the *Manchester Guardian* filed a report for his newspaper and *The Times*, prefixing it with the comment that he hoped that his 'lame account may serve to hint now at the unshakeable valour with which they faced strange new terrors and agonies.'

Just under three weeks later the 2nd Coldstream were to suffer what was probably their grimmest Christmas of the Regiment's long and illustrious history. Eisenhower, the Allied Commander, was determined that the enemy must be given no respite and ordered another attempt be made to reach Tunis. By now the winter rains had come, which aggravated resupply and made Allied air operations, which were, in contrast to the Axis airfields, largely based on earth airstrips, while the Axis had a number of tarmac airfields, that much more difficult. Nonetheless the effort had to be made, and the first task was to recapture Longstop Hill, which dominated the route forward from Medjez. It had got its cricketing name from the earlier advance by the East Surreys, who, recognizing its tactical importance, had left a platoon here. Now it was the Coldstream who were tasked with retaking it. They were positioned near Medjez station, and on the morning of 22 December the Commanding Officer, Lieutenant-Colonel Stewart-Brown gave his orders to 'the usual order group – company commanders, specialist platoon commanders, RA officers, tank commanders and the American combat team second-in-command'. The plan was for an attack that night, and, having secured the hill, to hand it over to an American regimental combat team at midnight. The orders lasted for some time 'until it was quite clear that everyone had the picture and plan quite clear in their mind'. In particular, much care had to be taken in tying up the details of the handover to the Americans, especially since it would be in unknown terrain and in the dark.

'During the afternoon after the conference the battalion became a bee-hive of activity. Although no one in the forward companies was allowed to move at all, behind the wood and the station area everyone was running about with orders and messages, DRs [Despatch Riders] were coming and going to and from Brigade and late in the afternoon the gunners began to move in with their 25 pounders. Hot meals were cooked and handed out, extra ammunition was made available where needed and various signal trucks arrived from Brigade to net in on the battalion wireless group. Extra manhandled sets were produced as it was quite clear that, owing to the shortage of tracks, the larger wirelesses in vehicles would not be able to get right up to the foot of the hill and battalion headquarters.'

Luckily the day remained fine, but the ground underfoot was very wet.

By last light everything was ready and once darkness fell the companies began to move off, the first two miles by road and then cross-country towards the objective. Battalion HQ set up in a white 'mosque like' building, and no sooner had they done so than fire was opened. It became clear that this was from mortars firing on pre-arranged defensive fire (DF) targets. Luckily the leading companies were already through these and suffered no casualties. The 25-pdrs then opened up. Surprise had been achieved and by 2300 hours No 1 Company reported that they were on the top of the hill, although they had had casualties, including their Company Commander very badly wounded in the neck. They were, however, having great difficulty in digging in among the rocks and the Germans were putting down heavy fire. The Commanding Officer wanted to send No 4 Company up to help, but radio contact had been lost and an officer had to be sent to find them. Midnight came, but there was no sign of the Americans. Indeed, it was not until 0230 hours that their leading elements began to appear. 'It was so dark by now that one could hardly see a thing more than ten yards away and it became obvious that the handover was to be ten times more difficult than anyone had anticipated.' Even so, it was successfully completed by first light and the Coldstreamers began a long march amid heavy rain to a rest area four miles west of Medjez.

'By ten o'clock on the 23rd we were hidden in a position four miles behind MEDJEZ in a gully thickly covered with scrub and olive trees. The battalion in fourteen hours had marched seven miles, nearly all of it across country, fought a night action in which they had gained the objective asked of them, handed over to the Americans and marched back eleven miles in pouring rain along roads and tracks only fit for tracked vehicles. In our gully everyone tried to rest as much as possible; the situation wasn't made any easier by the rain which continued to pour down in cloudbursts at a time. We had nothing more than the proverbial groundsheets and it wasn't long before blankets and the few dry articles of clothing were soaked right through.'

The Americans, however, were soon in trouble on the hill, and at midday orders were given for No 4 Company to return to it immediately. They had hardly finished their breakfasts, let alone had the chance to catch up on sleep. That evening, shortly after dark, the remainder of the Battalion was also ordered back to Longstop. This time, though, there was transport to take them part of the way. They took up position on the western slope and remained there throughout the daylight hours of the 23rd. They mounted another night attack and regained the peak of Longstop, but not without casualties, which were especially heavy among the company officers and NCOs. It was only then, however, that they realized what the problem was. They had attacked a false peak, one which was dominated by another behind. One company now attempted to take this, but the Germans proved too strong.

Dawn on Christmas Eve found the Coldstream in a difficult position. Not only were they overlooked and under persistent mortar fire, but it was difficult to construct decent defensive positions amid the rock, but

'no wheels could get within 5000 yds of the forward companies had no vehicles of any sort within 3000 yds. Ammunition therefore had to be brought to the col by carrier and thence by hand, while casualties, which were occurring all the time, had to be evacuated by the same painful process. It is not therefore surprising that by dawn the remnants of No 2 Company, who were manhandling the loads, were completely exhausted, and to make matters worse the whole of LONGSTOP HILL was covered with sopping wet knee-high rosemary and heather.'

Communications, too, were very difficult. Even so, the Battalion and its American comrades clung on, and that night reinforcements arrived in the shape of a French native infantry company. Dawn on Christmas Day, however, revealed the beginnings of a major German attack. Mortar and artillery fire intensified and the French, who were equipped only with rifles, withdrew. By 1000 hours it became clear that, with mounting casualties and ammunition shortages, Longstop could no longer be held. Consequently the Coldstream and Americans withdrew, covered by the 3rd Grenadiers, who were brought up for this purpose.

Longstop cost the Coldstream 200 killed and wounded, including the Commanding Officer and Adjutant wounded; few company officers were still on their feet, and only one CSM and one of the original platoon sergeants remained. Yet the Battalion kept its cohesion and fighting spirit throughout. As for Longstop, it would not be the last time that its slopes would be the scene of intense combat.

Thereafter the Allies went on to the defensive, holding a long front with too few troops. Indeed, while operations were conducted by the British First Army, it in truth consisted of only one infantry and one armoured division until the end of January. This could not, for security reasons, be revealed to the British public and there was understandable puzzlement as to why First Army had made so little progress, especially when compared to Eighth Army, which was streaming through Libya and, indeed, entered Tripoli on 13 January.

In January a fourth infantry brigade, 38th Irish, arrived, and, at the end of the month, 46th North Midland Division. The Irish arrived at a time when the emphasis on both sides was on patrolling. They found themselves opposed by the crack troops of the German 5th Parachute Regiment and learnt much from the experience. As the 2nd London Irish Rifles put it:

'The Goubellat Plain became the war nursery. Here they first heard the zip of the bullet, the quick stutter of the Schmeisser, the whine of the shell, followed by his bark, and the bloodsome crump of the mortar, the most formidable of them all. All fired with intent to kill. Every week saw improved techniques in fighting and several spirited encounters took place on enemy positions and farms in no-man's-land. Captain J. Grant took a party of men well behind the German lines and raided a German HQ, but Captain Grant and some of his men were wounded and taken prisoner. German prisoners said, "You are brave but not very good". This was fair criticism. Forty-eight-hour long-distance patrols paid a small dividend and were discontinued. The Arabs were usually giving away the daytime hide-outs. During this time the weather was extremely bad. Troops in their trenches were up to their knees in water. Roads were mud-tracks, quite unfit for men or vehicles.'

During January the Germans did launch a number of spoiling attacks, but in mid-February they mounted a larger operation, the brunt of which fell on the Americans, who suffered severely around Kasserine. This was eventually halted, but the Axis attacks in western Tunisia continued into the second half of March.

Meanwhile Montgomery, having halted at Tripoli in order to build up supplies, crossed into Tunisia. His first major brush with the enemy came on 6 March at Medenine, when Rommel, frustrated in the west, turned on him. Forewarned by Ultra, Montgomery was able to prepare a strong anti-armour defence and the German tanks were rebuffed with heavy casualties. The bulk of the German effort came up against 131 Brigade and 201 Guards Brigade, and the anti-tank guns of the 1st/6th and 1st/7th Queens particularly distinguished themselves. Now the Eighth Army closed up to the Mareth Line, but first had to clear the enemy from some hills known as The Horseshoe, from their shape. Montgomery gave this task to 201 Guards Brigade, promising them 'a good party'. Sadly, it turned out to be anything but. There was little opportunity for detailed reconnaissance and when the 3rd Coldstream and 6th Grenadiers attacked on the night 16/17 March they suffered grievously from mines, whose presence had been unknown beforehand. The objectives were taken, but the companies were too weak to hold them in the face of the inevitable German counter-attack and the mines prevented them from being reinforced. The Coldstream suffered 159 casualties and the Grenadiers, considering that they were a motor battalion, with only three rifle companies, a staggering 363. Indeed, so closely laid was the minefield that no less than 720 mines had to be lifted before the bodies of sixty-nine Grenadiers could be recovered for burial.

The forcing of the Mareth Line itself was spearheaded by the two* brigades of 50th Division, 69 and 151 Brigades, in a night attack. The lesson of The Horseshoe had been learnt and careful reconnaissance was carried out. Sixty-Ninth Brigade had also developed a system for securing lanes through the minefields. Each battalion had a picked fighting patrol, known as 'The Thugs'. These, accompanied by Sappers and equipped with scaling ladders to overcome the anti-tank ditch in front of the Axis defences, advanced ahead of the main body. Once the Sappers had cleared a lane the Thugs rushed in and attacked all nearby enemy posts so that their battalions could pass through without incurring too many casualties. This technique worked extremely well and enabled the Brigade to secure its objective, the Bastion. During the attack Lieutenant-Colonel 'Bunny' Seagrim of the 7th Green Howards won the Victoria Cross for personally leading assaults on a number of machine-gun posts. Sadly, he was to be killed two weeks later at Enfidaville before the award was announced. It was, however, 151 Brigade, consisting of three battalions of the Durham Light Infantry, which was to bear the brunt. It was commanded by Brigadier Daniel Beak, who had won four decorations for bravery, including the Victoria Cross, during 1914–18 while serving with the 63rd (Royal Naval) Division and had taken command just after Alamein, relinquishing the rank of Major-General which he had held while commanding the troops on Malta. Simultaneously with the assault on the Bastion, the 8th and 9th Durhams were to break through the Mareth Line and establish a bridgehead. Their main obstacle was the Wadi Zigzaou and in front of this lay a minefield, which was to be breached by Scorpion flail tanks of 41st Royal Tank Regiment.

This operation was successful, but went more slowly than planned. The Durhams then tackled the wadi, having at one point to form a human chain in order to get out of it. They seized their initial objectives, albeit

---

* 150 Brigade, which had been lost at Gazala, had not been reformed.

suffering heavy casualties. Valentine tanks of 50th Royal Tank Regiment now began to move forward across the wadi and drop fascines in the anti-tank ditch beyond so that they could cross over it. This also took time, but four tanks managed to get over before the fifth stuck and could not be cleared before first light. This meant that the Durhams were now isolated and had to endure a day of heavy fire. That night they resumed their attacks, with the 6th Durhams and 5th East Yorks (loaned from 69 Brigade) also being brought into the fray, while another crossing point was eventually created and the remainder of 50 RTR crossed.

At midday on the following day, just as the crossing point was being refurbished so that it could take wheeled vehicles, there was a sudden downpour. This meant that the Durhams' 6-pdr anti-tank guns could not be got across. That afternoon the bridgehead was attacked by the whole of 15th Panzer Division. The Valentines, most of which only mounted 2-pdrs, were outgunned and, after losing half their number, fell back to the anti-tank ditch. The Durhams had some of their positions overrun and were forced, after darkness had fallen, to withdraw to the anti-tank ditch. The Officer Commanding C Company of the 8th Durhams recalled that they had to leave their badly wounded behind when they fell back to the ditch.

Among them was a Lance-Corporal Bainbridge, who had been captured at Arras in 1940; he escaped from his POW camp in Germany and, with another 8 DLI man, walked through Poland into Russia where they were arrested as spies. After a year in Russia he was repatriated. He was sent to the Middle East as a reinforcement and rejoined the Battalion after Alamein. And now he was put into the bag again.

The remnants of all three Durham battalions were now intermingled. Plans had been hastily made for 69 Brigade to attack that night in order to restore the situation, but then Montgomery changed his plan, switching to an outflanking movement through the Tebaga Gap, which eventually forced an Axis withdrawal. The remains of 151 Brigade were therefore withdrawn, having suffered some 700 casualties. For those who had taken part it was a bitter blow, which was compounded later when Montgomery sacked both 50th Division's commander and Brigadier Beak, but, by tying down the Axis armour, the Division did enable the blow through the Tebaga Gap to succeed.

The Eighth Army had two more significant battles, again against defensive lines. At Wadi Akarit 4th Indian Division, led by the 1st/2nd Gurkhas and 1st Royal Sussex, carried out a successful silent night attack before the main assault, by 51st Highland Division and the hard-worked 69 Brigade, went in at dawn. They broke through and withstood counter-attack by the Axis armour before the Italo-German forces once more withdrew northwards. At Enfidaville, on the other hand, Montgomery met with a rebuff from a position which was just too tough.

By this time the First and Eighth Armies had joined hands, and the differences between the two became marked. The one was flushed with the success of an advance that had taken it almost 2,000 miles, while the other had spent a grim winter in the Tunisian hills. The Eighth's vehicles still retained their desert hue, while the First's were all painted green. Montgomery's men resembled Wellington's in their varied dress, while the First Army closely observed dress regulations. The Eighth Army regarded the First Army as amateurs, but as General Sir Gordon Macmillan, who served as Chief of Staff to both IX Corps (First Army) and X Corps (Eighth), said of the Eighth Army:

'They had high morale but little tactical sense. Their experiences in the Desert had taught them that the only way to attack the enemy was to advance against him under a creeping barrage. The result was that when they got into broken country with woods and obstacles, they had little elementary tactical sense and it was clear that they had forgotten much of what they must have learnt when they started their training in UK about ordinary platoon tactics of fire and movement.'

There was, however, one item possessed by First Army which was envied by the Eighth – their food. Montgomery's men subsisted on much the same ration as their fathers had during 1914–18. The only additions were occasional issues of tinned fruit and potatoes, and soya links (usually masquerading as sausages). The First Army, on the other hand, had a new field ration, Compo. John Horsfall of the 1st Royal Irish Fusiliers called it 'one of the most inspired rations ever served up to the British infantry' and described the food in a letter home as 'all tinned of course and includes bacon, sausage, cheese, butter, biscuits, tea mixture, puddings, and things like steak, Irish stew, Maconachie and many other delicacies, so we don't starve'. In other words, Compo had much greater variety and was issued in 24-hour packs, either individual or for five men, and consisted of food for three meals, sweets, cigarettes, and even lavatory paper. Although some of the menus have changed, Compo remains the British Army's staple ration.

In spite of Eighth Army's efforts to force a decisive conclusion to the campaign in Tunisia, Alexander, commanding the Allied ground forces, decided after Enfidaville that the prospects were more favourable in First Army's sector, and 7th Armoured and 4th Indian Divisions were transferred from Eighth Army. First Army had also received further reinforcements from Britain. The 1st Infantry Division had arrived in mid-March and 4th Mixed Division in early April. This had two infantry brigades and an infantry tank brigade, the object being to improve infantry-tank co-operation by giving the infantry division its own integral armour. Five divisions, the others being 1st, 3rd, 43rd and 53rd Divisions, adopted this organization, but it was found to be unsatisfactory because there were not sufficient infantry reserves and the concept was abandoned after the end of the Tunisian campaign.

First Army had already gone over to the offensive with an operation launched on 7 April and designed to secure the high ground dominating the road from Oued Zarga to Medjez el Bab. The Tunisian veterans, 78th Division, were given this task and, in ten exhausting days of fighting over terrain which the army commander described as 'a kind of Dartmoor or Central Sutherlandshire, but with deeper valleys and steeper hills', they achieved their objectives and Medjez could once more be used as a springboard for a thrust on Tunis. The next significant objective was the infamous Longstop Hill, which was assaulted by 36 Brigade on the night 22/23 April. The 5th Buffs managed to secure their initial objective, but the 6th Royal West Kents were badly held up by minefields. Consequently the 8th Argylls were committed next morning, supported by an artillery barrage, which included much smoke on certain features, and a troop of Churchill tanks of the North Irish Horse. Lieutenant-Colonel Taylor, who was then Second-in-Command of the Battalion, takes up the story:

'The advance was across wheat fields almost ready to harvest, and the companies were in very extended order to try and minimize casualties. After a month's hard fighting the coy strengths were low and rifle coys were about 50 strong. Together with the various support, signals, and intelligence, etc, we were about 300 strong on the start line.'

'I watched until the leading troops reached the first ridge on the lower slopes and went to Bde HQ to give a verbal report to the Brigadier. I knew that casualties were high, but at this stage I was certain that the hill would be taken. In the late afternoon I received a message that the CO [Lt Col C. V. O'N McNabb, who had, at his own request reverted from Brigadier and Chief Staff Officer of First Army to take over the 8th Argylls] was missing and thought to have been killed and I proceeded to Longstop, accompanied by Major Alec Malcolm (Support Company Commander) who had earlier been on the hill. We travelled by Bren carrier to the foot of the Hill and then we set out on the stiff climb, each carrying a 5-gallon can of water. The day was very hot and I judged that water would be the first priority. On the way we met up with one of the NIH tanks which had succeeded in climbing over the rocks and was only a short way behind the leading troops. It was a comforting feeling to know we had their support so close up. It had its disadvantages, of course, as a tank always attracted lots of artillery fire.'

'When I reached our forward position I found a force of 5 officers and about 40 men. These were the men who had succeeded in capturing the final peak. By nightfall we gathered up some very exhausted stragglers and were able to muster about 100 all ranks.'

Major Jack Anderson DSO had led the final attack on the top, in which he and several of the men with him were wounded. He was awarded the Victoria Cross, but was killed at Termoli in Italy in October, 1943. The 1st East Surreys came up that night to reinforce the Argylls and after two days' further fighting the Djebel el Rhar feature, which had given the Coldstream so much trouble four months before, was seized. During this time Lieutenant-Colonel Wilberforce of the East Surreys was the only commanding officer in the Brigade still on his feet. Sadly, he, too, was killed, by an artillery shell, one of the last rounds fired in anger at the East Surreys in Tunisia, on 6 May. The performance of 78th Division as a whole during the past two weeks won high praise from General Anderson, the army commander, who considered its operations 'as tough and pro-longed a bit of fighting as has ever been undertaken by the British soldier'.

While Longstop was being fought for, 1st Division had become involved, securing the high ground on the south side of the Medjez-Tunis road. Further south 4th Division found itself embroiled in a bitter fight at Peter's Corner against the crack paratroopers of the Hermann Goering Division. First to suffer were the 2nd Royal Fusiliers, who had their Commanding Officer killed by a mortar bomb. Lieutenant-Colonel M. L. Brandon had been RSM of the 1st Battalion for almost nine years when he was promoted to be Quartermaster of the 2nd Battalion in May, 1939. After distinguish-ing himself during the retreat to Dunkirk he had been granted a combatant commission and within two years had risen to command the Battalion. He was, as the Battalion's war history put it, 'a born leader of men, a man of strong character but with a delightful manner, and he was greatly liked and deeply respected by officers and men alike. Quite fearless, he inspired all ranks with the greatest confidence.' The two other battalions in 12 Brigade, 6th Black Watch and 1st Royal West Kents, also suffered, as did the 2nd Duke of Cornwall's Light Infantry from 10 Brigade. They were eventually relieved by 11 Brigade.

The final important feature barring the way to the Tunis plain was Bou Aoukaz, which was attacked by 24 Guards Brigade on the afternoon of 27 April. As at Peter's Corner, the Germans clung on tenaciously, and in the ten days' battle that followed the Brigade won two Victoria Crosses, posthumously to Captain Lord Lyell of the 1st Scots Guards and also to

Lance Corporal Kenneally of the 1st Irish Guards. One who took part recalled:

> 'During this period it was difficult to keep abreast with time. Day followed night, and we had no idea which particular one it was. The weather was becoming hot, and we were still clad in battle dress. Flies were abundant, and those of us left were suffering from severe diarrhoea and in some cases, dysentery. Meals were unheard of – sometimes we might go for 2 days with no food or water. The best way to keep saliva in the mouth was to suck a pebble. Ammunition had priority over food and water.'

Eventually the Guards were relieved, each battalion reduced to a strength of less than forty men, by 2 Brigade and the Bou finally fell.

Finally the armour could be released and it was symbolic that both First and Eighth Armies were equally represented in the last thrust on Tunis. With 4th Mixed and 4th Indian Divisions to secure the start line, 6th and 7th Armoured Divisions were loosed from their leash, patrols of their two armoured car regiments entering the city at much the same time on 7 May. After some mopping up Alexander was, on 13 May, able to send his famous cable to the Prime Minister: 'Sir: It is my duty to report that the Tunisian campaign is over. All enemy resistance has ceased. We are masters of the North African shores.'

Historians and others have been prone to regard Tunisia as a mere afterword to the cut and thrust of the Desert Campaign. Likewise, the part played by First Army tends to be mistakenly overshadowed by that of the Eighth, and this was not helped by the fact that no sooner had the campaign finished than it ceased to exist, with Eighth Army taking over all its troops. The truth of the matter is that Tunisia was a very tough campaign and, in contrast to the Desert, essentially an infantryman's war. As such, it demanded many of the traditional qualities of the British infantryman – doggedness, pluck and determination. The experience gained and the lessons learnt would stand them in good stead for the Allied re-entry into the European continent.

There was a pause after the end of hostilities in North Africa, one that enabled battalions to be brought back up to strength and for the winter mud and spring dust to be washed away in the sparkling waters of the Mediterranean. In the meantime, planning for the Anglo-American invasion of Sicily, which had begun while the fighting continued in Tunisia, was finalized. The infantry divisions selected to take part in the initial assault were 1st Canadian, which had spent over three years training in England, 50th Northumbrian and 51st Highland, now beginning to establish a reputation as shock divisions in Montgomery's eyes, and 5th Division. The last-named had led a nomadic existence during the past year. After two of its brigades had been involved in the Madagascar campaign, the Division had gone to India and then to Persia, Iraq and Syria. Now it was finally to have the opportunity, the first since May, 1940, to fight as a division. The landings themselves, spearheaded by Commandos, which was now becoming standard practice, were successfully achieved on 10 July and beachheads quickly secured in the face of low grade Italian coastal divisions. The Sicilian terrain and the fact that the island's reserve consisted of two good German divisions, 1st Parachute and 29th Panzer Grenadier, meant that the campaign would be no walkover. The divisions also landed on light scales, which meant that vehicles were limited and most movement was done on foot. Some idea of the problems is given by Colonel H. B. L. Smith MC, then commanding the 1st East Surreys, who landed as part of 78th Division in the follow-up behind the initial landings:

'From thence to Bronte and Randazzo the advance was restricted by the only road around the slopes of Etna with often a sheer drop on one side to the Sineto river; on the other side lava outcrops made movement painful and difficult. Armour could only take its turn on the road; and much of the time we were beyond our artillery support, the guns being unable to deploy off the road. It was here we first encountered the enemy "wailing minnies" or Nebelwerfers – six barrelled rocket mortar, fired electrically. The advance was made hideous by their noise together with the explosions of the German demolition parties.'

Eventually, though, the twin drives by Patton's Seventh US Army in the west and Montgomery's Eighth in the east drove the Axis forces back towards the north-east corner of the island, and, with the evacuation of their remnants across the Straits of Messina, Sicily was secured.

Two campaigns in semi-mountainous country had brought out several valuable lessons as far as the infantry were concerned. Colonel Smith again: 'Amongst other things we had found that with determination and good leadership throughout the Bn we could do quite a lot despite lack of food and very difficult and strenuous physical conditions.' Improvisation was often a must, as was 'the old lesson that you can only be sure of having to hand in battle such weapons, tools, food and water as you are able to carry on the man'. There were other lessons, too, which would be equally applicable to much of the fighting in Italy:

'Mules and porters will always be vital over some types of rocky terrain which cannot be negotiated by tracks or wheels, and their needs should be taken into account when planning . . . We sometimes used to laugh at the pre-war training as practiced on the NW Frontier of India and yet here again we were using that old machine gunner's tool, the Pointer Staff, to indicate obscure features so often impossible to pick out quickly on inaccurate maps. Once again we found ourselves picquetting the heights with communications breaking down. Sema-phore, helio or a knowledge of Morse would have filled the gap. The wireless sets we had then were not robust enough to stand up to rough treatment on rocky terrain and they were too often subject to screening or interference by mountain features . . .'

'After too many painful experiences of finding ourselves overlooked by the enemy on a forward slope, we now went for the tops or crowned the heights, using defilade and reverse slopes for our better protection. We had to relearn how to make a sangar, as digging on the rocky terrain so often was almost impossible . . .'

'Finally we had learned a very healthy respect for the enemy.'

The Italian mainland now beckoned, and on 3 September XII Corps crossed the Straits of Messina and landed in the toe of Italy. Six days later came the main landings at Salerno, with a subsidiary operation by 1st Airborne Division against the port of Taranto. Although the Italians had now sued for an armistice, the Germans had speedily moved into southern Italy, and Salerno showed that they still had plenty of fight even though they had been abandoned by their ally. The landings were conducted under Mark Clark's US Fifth Army and were carried out by two American and two British divisions. The all Hampshire Regiment 128 Brigade led 46th Division and 167 Brigade and the all Queen's Regiment 169 Brigade of 56th Division. The feelings of one who took part in the landings, Private Blay of the 2nd/6th Queens, as the armada approached the shore:

'Eventually came the order to get dressed and stand by. Pack on – check rifle, one up the spout, safety catch on. Right, up on the deck said the sergeant. Nervously I followed the men in front up on the deck. Gunfire could be heard on our right. I could see tracer in the air and shell bursts. In that direction also seemed to be an orange glow. I could hear machine-gun fire. I thought shall I make it to the shore? My greatest dread was would I be shot and fall before we reached dry land . . .'

'We are coming in fast now and then there is a great grinding as the bottom of the craft charges onto the beach. Down go the ramps on either side. Ashore and no injuries, my greatest worry over, and we are shepherded up the beach by the powers that be.'

Fortunes varied among the battalions on landing. Few, though, landed in precisely the right place and a certain amount of adjustment of positions was needed. The Queens Brigade secured all its initial objectives, but did have casualties, notably among the 2nd/5th, who had over 170. It was, however, the Hampshire Brigade which suffered most, especially the follow-up 5th Battalion, which took time to sort itself out on the beach. By the time it began to advance inland it had become separated from its anti-tank guns and its PIATs were not in working order. As the Hampshires moved forward on three tracks, taking few tactical precautions, they suddenly collided without about 100 Panzer Grenadiers, supported by tanks. The result was virtually a massacre. Within a few minutes the Hampshires had suffered about 100 casualties and a further 300 surrendered, with only some six officers and thirty men getting back to the beachhead perimeter. It was the second time that this battalion had suffered badly at the hands of tanks, the first being against Tigers at Sidi Nsir in Tunisia the previous February. The 2nd Hampshires, heroes of Tebourba, also became involved and, again being without their anti-tank guns, lost two rifle companies. All this contributed to delays in exploiting inland and gave the Germans more time to build up their forces. In the meantime the follow-up brigades were landed.

The second day produced continuous attacks by the Germans, but they failed to break through, even though there were too few troops holding the perimeter, with some battalions being totally unsupported. One was the 9th Royal Fusiliers, who had 300 men captured, including their commanding officer. The pattern of German attack, followed by counter-attack, continued during the next few days until the Germans had shot their bolt. Only then was Fifth Army able to advance, having linked up with Eighth Army coming up from the south on the 16th, pushing through the mountains to enter Naples on 1 October. Thereafter Kesselring conducted a skilful withdrawal to the first of his major defence lines, which would so dominate the fighting in Italy, the Gustav Line, running across the breadth of the country some 75 miles south of Rome. This was to hold the Allies until well into the new year.

One sideshow mounted in the Mediterranean in Autumn, 1943, brought disaster to five British battalions. Churchill was keen to exploit the vacuum in the Aegean created by the Italian surrender, especially by occupying Rhodes. Here the Italians surrendered to the Germans, but, undeterred, 234 Infantry Brigade, together with elements of the RAF Regiment and some Paratroopers, were rushed by destroyer to Leros, Samos and Cos.

The Germans quickly woke up to what was happening and began to attack from the air. They landed on Cos, which was held by the 1st Durham Light Infantry and an RAF Regiment unit, who were soon overwhelmed. Only about a hundred men were eventually rescued by caiques of the Special Boat Service. The Combined Chiefs of Staff refused the air assets necessary to ensure that the other two garrisons could hold out and in mid-November they, too, were overrun. A mere 250 men of the battalions involved – 4th Buffs, 2nd Royal Irish Fusiliers, 1st King's Own and 2nd Royal West Kents – managed to escape.

On 16 September, 1943, the day on which the two Allied armies linked up after the Italian landings, occurred a distressing incident in the British sector of the Salerno beachhead. Some 700 infantry replacements, just arrived from North Africa, refused to join frontline units and staged a sit-down. It transpired that all came from the 50th Northumbrian and 51st Highland Divisions, which were then still in Sicily, and that many were veterans of the fighting in the Desert and Tunisia and had been sick or wounded. Not until they were at sea were they informed of their destination, having assumed that they were bound for Sicily. Their disgruntlement at being sent to different divisions in a strange army, with Highlanders being posted to English county battalions, is understandable. Their anger was possibly also compounded by rumours that 50th and 51st were to be sent back to Britain to prepare for the long-awaited opening of the Second Front. Certainly Montgomery was proposing to Alan Brooke that this should happen while Salerno was being fought, and by early October he was warning Alexander:

'We are playing with a highly explosive material when dealing with the question of formations and personnel going to the UK . . .'

'50 and 51 Divs know they are going home to UK, they have had their sick and wounded, when fit, sent to Divisions in Fifth Army.'

'This may have been necessary as an emergency measure, but it is vital that these men be sent back to their Divisions at once – to avoid unrest.'

In the meantime General Dick McCreery, the British Corps Commander at Salerno, had visited the mutineers and promised them that they would be posted back to their own units as soon as possible and their disobedience forgotten, provided they obeyed orders now. The majority agreed to this, but 192 men held out and, with remarkable insensitivity, they were put in a compound next door to one containing German prisoners, who jeered at them and called them cowards. They were then shipped back to Constantine in North Africa and court-martialled for mutiny. All bar one were found guilty and sentenced to either seven or ten years' hard labour, dependent on rank. The three sergeants in the group received the death sentence. Suddenly, possibly as a result of Montgomery's agitation, the sentences were suspended provided that the men agreed to soldier with to whatever unit they were now posted. This they did.

The root of the problem was a weakness in the regimental system. While manpower was plentiful it could be strictly observed in wartime, but by this stage of the war the barrel was beginning to run low. Indicators of this had already been seen in Tunisia. One of the first was the dissolution of 44th (Home Counties) Division in January, 1943, although its brigades and battalions survived, being sent to other divisions. At the end of that campaign a number of battalions were disbanded, their men not necessarily being sent to sister battalions of the same regiment who were in the theatre. Thus the 2nd North Staffords received a draft of eight officers and 172 men of the 1st/6th East Surreys before May, 1943, was out. Luckily they were not posted to Italy until the end of the year and there was time to ensure that the new arrivals were properly assimilated.

The problem was, however, no different from that of 1914–18, when there was probably even less regard for the niceties of the regimental system. Being posted to a new regiment and made to change capbadge was probably resented more during the 1939–45 because the solider was that much more articulate and had higher expectations than Old Bill and Young Bert had had in the trenches. Indeed, the Army was very conscious of this and in 1941 published a pamphlet entitled *The Soldier's Welfare: Notes for Officers*. This made it very clear that:

'Discontent seldom arises from hardship, provided that the men feel that hardship is reasonable, ie, that it is a necessary part of the business of winning the war. They are ready to endure cheerfully anything which they believe to be unavoidable, but they are easily disgruntled if they feel that the hardships are caused by red tape or by inefficiency . . .'

'Whenever possible, therefore, the reason for irksome orders or restrictions should be explained to him.'

This had clearly not happened in the case of the Salerno Mutineers, as Montgomery undoubtedly recognized, and so did McCreery, who later assumed command of the Eighth Army and commented that 'it was like an old steeplechaser, good for one more race if it is carefully handled'. From now on the British commanders in Italy, and elsewhere, had the preservation of life as a high priority, because of the growing manpower shortage, and relied increasingly on firepower and sound tactics to achieve their objectives. This was in contrast to the Americans in Italy under Mark Clark, to whom casualties did not matter so much, at least until Autumn, 1944, and who often accused their British colleagues of insufficient drive.

Nonetheless, the Eighth Army in Italy considered itself a victorious army during the latter stages of and after the North African campaign. In the India/Burma theatre the challenge of instilling self-belief in a defeated army was very much greater.

CHAPTER SEVEN

The Jungle is Neutral
Burma 1942–3

THE overrunning of Malaya, Singapore and Burma totally reversed the British view of the Japanese soldier. From being an inferior individual in every respect he was now regarded as a superman, with the jungle as his friend. For those who had survived the encounter with the Japanese in Burma, their reception in India had forced their morale down still further. Luckily in Slim the troops had a commander who was equal to the task of restoring both morale and combat efficiency. Not only had he supervised the final retreat from Burma, but he also had extensive battle experience. This had begun with Gallipoli, as a Kitchener volunteer in the Royal Warwicks and where he had been severely wounded, and subsequently in Mesopotamia, where he was wounded again. He had then served on the North-West Frontier with the 6th Gurkhas, to whom he transferred in 1920, having admired at first hand the 1st/6th's gallant performance in the Dardanelles in 1915. He had then commanded an Indian brigade in East Africa during the early 1940 skirmishes with the Italians, and command of 10th Indian Division during the crushing of the Iraqi revolt in summer 1941.

Slim was given command of XV Corps in Bengal and Orissa in northeast India. Among the formations under his command was 70th Division, which had been sent to India from the Middle East in early 1942. Most of his troops were involved in internal security duties, caused by Gandhi's civil disobedience campaign and fuelled by a severe famine in Bengal, and it was only once order had been restored that XV Corps could begin serious training. Slim first moved his men away from the fleshpots of Calcutta to the Ranchi plain, which contained large tracts of jungle. His training programme was built around eight lessons that he himself had learnt during the first campaign in Burma. For a start, the soldier must learn to regard the jungle as 'neither impenetrable nor unfriendly' and to use it to his advantage. Patrolling was 'the master key to jungle fighting', and every soldier, not just the infantryman, had to learn to patrol 'boldly, widely, cunningly, and offensively'. Units had to become used to the Japanese operating in their rear, and the concept of linear defence had to be eradicated. Penetrations had to be quickly dealt with by mobile reserves. Frontal attacks were not to be practised; instead, the enemy should be tied in the front and attacks launched in the flank or rear. Tanks were able to operate anywhere except in swamps, but co-operation with infantry had to be the very closest; 'penny-packeting' of armour was unacceptable. There was no such individual as a non-combatant in the jungle; every type of unit had to be prepared to protect itself. Finally, it was crucial to 'regain and keep the initiative'. As 1942 wore on Slim's men, living in the jungle for weeks at a time, grew in skill and confidence, but this did not apply to the whole of the army in India.

Many battalions still found that aid to the civil power gave them little opportunity for field training. The experience of the 9th Border Regiment is typical. They landed in India in Spring, 1942, and were sent to Calcutta. Not only did they find themselves having to provide guards for vulnerable points and help quell riots, but often had to supply escorts prepared to travel all over India and even to Ceylon. 'It was hard and thankless work, and often a Lance-Corporal and a couple of men were away on responsible jobs for weeks at a time with few provisions for their travel and with food difficult and every expensive to get.' While plenty of social amenities existed for the officers, there were still few for the men, and the Battalion was more than relieved when it was finally able to begin serious jungle training.

In later Summer, 1942, plans were set in motion for a limited offensive in the Arakan, designed primarily to recapture the island of Akyab in the Indian Ocean. The task was given to General Irwin's Eastern Army, who in turn tasked 14th Indian Division. In the event, lack of air support and amphibious shipping forced the cancellation of a direct thrust against Akyab and an advance down the Mayu peninsula and the east bank of the River Mayu was substituted, with the idea of isolating Akyab. Irwin wanted to improve his tenuous lines of communication to Cox's Bazaar, his launching pad. This included the construction of a road, but this was delayed by heavy rains, and it was not until just before Christmas that the advance really got underway. This was at the beginning of the dry season, but even so, Arakan, while in many ways very attractive, is highly malarial and a paradise for leeches because of the very high rainfall during the wet season, which lasts from June until October. The advance initially went well against light opposition, but the Japanese, realizing the threat to Akyab, began to reinforce it and its land approaches. East of the Mayu, 123 Indian Brigade reached Rathedaung on 28 December, but could make little impression on it, its British battalion, the 10th Lancashire Fusiliers, suffering heavy casualties in two attacks on the town. Likewise, 47 Indian

Brigade on the Mayu peninsula was halted at Donbaik. Here the Japanese had constructed shell-proof bunkers made of earth and logs and the 1st Royal Inniskilling Fusiliers made a number of attacks against them, but were foiled each time. Further troops were thrown in, but the Japanese continued to resist. The number of brigades committed to 14th Division's command eventually rose to ten, nigh on impossible for one headquarters to control, and the climax came towards the end of March.

The British 6 Infantry Brigade, part of 2nd Infantry Division, which had arrived in India in Spring, 1942, had been strengthened to six battalions through attachments. It was ordered to attack Donbaik, but failed, like previous attempts. This was in spite of particular dash exhibited by the 1st Royal Welch Fusiliers. Meanwhile the Japanese 55th Division was moving westwards and behind 14th Division's forward brigades. Valiant efforts were made by 47 Indian Brigade to keep them at bay, but without success, and eventually the Brigade was forced to break up into small groups and make for the coast. The Japanese then struck 6 Brigade, overrunning the headquarters and capturing the Brigade Commander and some of his staff, most of whom were later executed, and inflicting casualties on the 1st Royal Scots. The Commanding Officer of the 2nd Durham Light Infantry took over the brigade and set about extricating it.

Slim now took personal charge, but his attempts to try and draw the Japanese into battle on ground of his own choosing failed, largely because of the poor state of the troops. Many of them had been fighting for over six months, short of food and wracked with malaria. They had become increasingly frustrated by their lack of progress and morale had plummeted. A liaison officer sent by Slim to visit 4 and 6 British and 71 Indian Brigades describes their state well:

'Outstanding was the fact that our troops were either exhausted, browned off or both, and that both British and Indian troops did not have their hearts in the campaign. The former were obviously scared of the Jap and generally demoralised by the nature of the campaign, ie the thick jungle and the subsequent blindness of movement, the multiple noises of the jungle at night, the terror stories of Jap brutality, the undermining influence of fever, and the mounting list of failures; the latter also fear the jungle, hate the country and see no object in fighting for it, and also have the feeling that they are taking part in a forgotten campaign in which no one in authority is taking any real interest.'

Thus withdrawal back into India became the only option.

The Arakan campaign did nothing to restore the morale of the troops in

India and merely reinforced the view of many that the Japanese were vastly superior to them in jungle warfare. There had, however, been another operation taking place at the same time which showed more promise. This was the first Chindit Expedition.

Much has been written about Orde Wingate, and he remains a highly controversial figure. He came to Burma with a background of unconventional warfare, having organized the Jewish Night Squads during the Arab Revolt and then commanded Gideon Force of Sudanese and Abyssinians, which accompanied the Emperor Haile Selassie during the liberation of Abyssinia. Wavell summoned him to Burma in early February, 1942, originally to coordinate the operations of the Chinese Fifth and Sixth Armies. This was changed and the American 'Vinegar Joe' Stilwell given the task instead. Wingate was now tasked with coordinating irregular warfare in Burma, with the aim of slowing down the Japanese advance. The forces available were two Special Service Detachments (SSD), which were part of Mission 204, originally set up to organize guerrilla operations in China. By the end of March Wingate was already using the term 'long range penetration' in his concept of operations for the SSDs, but events moved too quickly for them to be put into effect and the surviving members were forced to cross into China before being repatriated to India. Once back in India Wingate developed the concept further, arguing that the low density of Japanese troops in Burma, together with the nature of the terrain, gave behind-the-lines operations, designed to disrupt the Japanese communications and keep them off balance, great potential. The one proviso was that the troops would be resupplied from the air. Wingate sold the idea to Wavell, who, in July, 1942, agreed that he could raise a force for this purpose.

Two battalions were initially made available for what became known as 77 Brigade. One was the 3rd/2nd Gurkhas, a recently raised battalion, only one of whose officers had so far seen active service. The other battalion was even more unusual. The 13th King's Liverpool Regiment had been sent to India primarily as a garrison battalion. Most of its men were over thirty and married; hardly the material for what Wingate had in mind. In addition there were survivors of the two SSDs, who were formed into 142 Commando Company, and the 2nd Burma Rifles to act as his reconnaissance element. Wingate's training doctrine was based on pushing men to the limit. As one of his officers, Mike Calvert, later wrote: 'If you have marched thirty miles in a day, you can take twenty-five in your stride.' For the King's, to be suddenly transferred to a jungle training area, invited to live in bashas, with few comforts, and to be as stretched as they were, it came as a shock. Inevitably, a number could not take it in, which was not surprising, and 250 men, including the Commanding Officer, were weeded out, their place being taken by younger men. The force was organized as columns, each based on a rifle company and capable of operating independently. A column's transport consisted of fifteen horses and 100 mules.

The original plan was for the Chindits, as Wingate dubbed his men, having misheard the Burmese word for lion (*chinthe*), to operate in conjunction with a main force operation. This, however, was not forthcoming and so Wingate persuaded Wavell that he should act on his own, on the grounds that his men's training had been geared towards an operation during Winter, 1942–43. The Chindits therefore crossed the Chindwin on the night of 13/14 February, 1943, tasked with sabotaging railway lines, cutting supplies to the two Japanese divisions in northern Burma, and general harassment. The story of the first Chindit expedition has been told many times. Suffice to say that in spite of losing two out of the seven columns early on, the Chindits succeeded in cutting the Myitkyina-Mandalay railway in a number of places and mounted some successful ambushes. More significant was that they were able to operate behind Japanese lines for some four months, totally reliant on resupply by air. Even so, almost a third of the 3,000 men who set out failed to return to India, and, because of the hardships, few of the survivors were ever fit enough to return to combat. Nevertheless, Wingate had proved that the British and Indian soldier could be just as effective in the jungle as the Japanese. This was in stark contrast to Arakan.

In October, 1943, Slim was given command of the newly created Fourteenth Army. Now he could set about putting into practice the ideas which he had developed while commanding XV Corps. At bottom it was morale that needed to be raised and Slim did this in a number of ways. First, the scourge of malaria had to be overcome. Preventative measures were one of the keys to this and research in India produced a number of medicines, the best known being mepacrine. Indeed, the incidence of malaria in a battalion becamse as good a measure of its quality as that for trench foot had been during the Great War. For those who succumbed, Slim instituted Malaria Forward Treatment Units, field hospitals situated a few miles behind the front line, which enabled men to return to their units on average after three weeks, rather than the six months it took previously when all casualties were evacuated back to India. Likewise, treatment as far forward as possible also applied to battle casualties. At the same time, seriously wounded were to be evacuated by aircraft rather than face an often long and uncomfortable journey by road and train.

Much effort was made to improve the standard of food and to make rest and reinforcement camps more cheerful and efficiently run. To overcome the 'Forgotten Army' syndrome, a theatre newspaper, *Seac*, from the acronym of South East Asia Command, was established and troupes from

ENSA (Entertainments National Service Association) encouraged to come out from Britain to entertain the troops. The delivery of mail, that vital morale factor, also became more efficient. Slim, too, adopted the Monty approach of constantly showing himself to his soldiers, encouraging his subordinates to do the same. He also was at great pains to make the individual soldier believe that he was playing an important part and was not merely an army number. During First Arakan battalion commanders had complained that individual reinforcement had arrived virtually untrained, 'many of whom according to the CO of the LF [Lancashire Fusiliers] had never seen a Bren gun'. Two training divisions, 14th and 34th Indian, were therefore established, and every soldier passed through one of these to perfect his jungle skills before being posted to a front line unit. Finally, in the front line itself Slim instituted a programme of aggressive minor operations. Battalions selected their best men to form fighting patrols and were encouraged to bring back trophies. Small attacks were mounted, ensuring always that they were carried out in overwhelming strength in order to ensure success. By early 1944 Fourteenth Army was beginning to believe in itself. As Slim wrote: 'We had laid the first of our intellectual foundations of morale; everyone knew we could defeat the Japanese, our object *was* attainable.'

CHAPTER EIGHT

Victory Road
North-West Europe 1944–1945

ON 2 January, 1944, Montgomery arrived back in England after sixteen months of highly successful campaigning at the head of the Eighth Army. His immediate task was to take command of 21st Army Group, formed for the cross-Channel invasion of France. As far as the initial British assault divisions were concerned, one had already been selected. This was the one remaining prewar Regular division still in Britain, 3rd Infantry. In March, 1943, it had been earmarked for the Sicily landings, but its place had been taken by 1st Canadian Division. With the setting up of 21st Army Group in July, 1943, the Division was one of the first to come under its command, and the divisional commander sought a personal assurance from Alan Brooke, the Chief of the Imperial General Staff (CIGS), that his division would be the first infantry ashore, and Montgomery, who had, of course, commanded it in the French campaign, was happy to uphold this. Indeed, the Division, stationed in the Scottish Lowlands, had already begun serious training in the late summer of 1943. That autumn it undertook amphibious training at the Combined Training Centre at Inverary, and in December moved to the Moray Firth, where the coast was similar to that where it would land in Normandy. Here it married up with Force S, the naval task force which would take it across the Channel, and began training with it.

Montgomery's first step on taking over 21st Army Group was to double the size of the initial landing sector, which meant an increase in the number of assault divisions. Thus, 3rd Canadian Division and those veterans of North Africa and Sicily, 50th Northumbrian Division, were selected as the British Second Army's contribution. The 50th Northumbrian had returned to Britain, together with 51st Highland and 7th Armoured Divisions, before Christmas, 1943, and had now been made up to three brigades by the inclusion of 231 Brigade. This had begun life as the Malta garrison, being originally called the Malta Brigade, and consisted of three Regular West Country battalions – 2nd Devons, 1st Hampshires, 1st Dorsets. In April, 1943, with the threat to Malta now evaporated, they moved to Egypt and became 231 Independent Brigade Group and, as such, took part in the

Sicily landings and those in the toe of Italy. Hence they had more experience of amphibious landings than any other British formation in 21st Army Group. But 50th Northumbrian also received another independent brigade, 56th, for the landings, but this was placed temporarily under command rather than becoming an integral part of the division. The remaining first line infantry divisions in Britain – 15th Scottish, 43rd Wessex, 53rd Welsh and 59th Staffordshire – also became part of 21st Army Group. The exception was 52nd Lowland Division, still configured as a mountain division, but its turn would come. All these divisions had received a leavening of veterans of the fighting in North Africa, Sicily and Italy and they could pass on the fruits of their experience.

Also included in the British order of battle were two airborne divisions, 1st and 6th, although the former, which had been through Tunisia and Sicily, was placed under First Allied Airborne Army. Each of these consisted of two parachute brigades and an airlanding brigade, which had three battalions who flew into battle in gliders. The airborne brigades, which comprised battalions of the Parachute Regiment, were formed from volunteers, although some units were created from the nucleus of existing infantry battalions.* In contrast, the airlanding battalions were ordinary infantry battalions, which had been converted to the role. They were organized as four rifle companies, which also possessed two 3-inch mortars, a support company with four 3-inch mortars and a reconnaissance platoon, pioneer platoon and administrative elements, and an anti-aircraft/anti-tank company with twelve 6-pdr anti-tank guns and twelve Hispano-Suiza 20mm anti-aircraft guns. Early in 1945, as a result of operational experience, the rifle companies were reduced from four platoons to three, the 3-inch mortars centralized, and the anti-tank guns reduced by four, with four Vickers machine guns being put in their place. In the case of 6th Airborne Division, which had a vital role on D-Day, the 6th Airlanding Brigade consisted of the 12th Devons, 2nd Oxfordshire and Buckinghamshire Light Infantry, who still liked to call themselves the 52nd, and the 1st Royal Ulster Rifles, none of whom had so far seen action during the war. Both they and their sister airlanding brigade in 1st Airborne Division would make good this omission before 1944 was out. The airlanding battalions were just one more indicator of the infantry's versatility.

Infantry battalions, too, were represented in 21st Army Group's British armoured divisions, with the exception of the 79th, which was the umbrella for specialized armour. Three conventional armoured divisions, 7th, 11th

* These were as follows:
  5 Para from Queen's Own Cameron Highlanders
  6 Para from 10th Royal Welch Fusiliers
  10 Para from 2nd Royal Sussex

and Guards, were to take part in the Normandy fighting, and each now consisted of an armoured brigade (three armoured regiments and a motor battalion) and a lorried infantry brigade (three battalions). The latter also had a machine-gun company, reflecting a change in policy towards machine-gun battalions. Now the idea was to give each infantry brigade a support company, thus reverting to the Western Front organization of the middle part of the Great War. In 1942 the machine-gun battalion itself had been augmented by a 4.2-inch mortar company, with sixteen mortars, and its title changed to infantry support battalion. The brigade support company consisted of twelve Vickers and four mortars. Divisional support companies were also introduced and were twice the size of those at brigade. Infantry formations also had available a number of independent armoured and tank brigades, the difference between the two being that the former were generally equipped with the Sherman and could strengthen the armoured divisions, while the latter had the slower but more heavily armoured Churchill, which were better able to give the infantry formations the close support that they would inevitably need.

As Spring approached the tempo of training increased. In wintry weather the landing exercises became even more of a challenge, but as General Sir Richard Goodwin, who was then commanding the 1st Suffolks in 3rd Division, recalled: 'I have never in my life seen troops so tough and fit. Despite the extreme cold and wet, their enthusiasm on these exercises was quite outstanding.' Others, though, became impatient and wanted to get it over with. In April 21st Army Group began to concentrate in southern England, the Americans in the south-west, and the British in the south-east. The final work-up exercises took place in early May under the codename FABIUS. These rehearsed a whole range of activities, which included for the infantry formations assault landings and a night advance through minefields. These completed, the assault troops then moved into camps in the marshalling areas close to the ports of embarkation.

On 26 May the final phase of the preparations were begun. The camps were sealed and, to all intents and purposes cut off from the outside world, the troops were finally told their specific tasks. For this purpose each camp contained specially set up briefing rooms, equipped with models of the relevant beaches, maps and air photographs. All place names were, however, disguised with codenames, and their true identity would not be revealed until overprinted maps were issued after embarkation. French francs and phrase books were issued, and then, a few days later, loading began.

One of the first to move was the 2nd Devons, who embarked on the *Glenroy*, one of the three famous Glen ships which had been in the amphibious operations business since 1940, at Southampton. They spent the next few days in Southampton water. On 4 June, however, the announcement came that the invasion was postponed for 24 hours because of the adverse weather conditions. Some, like the 2nd Royal Ulster Rifles, were lucky in that they had not already embarked and were able to return to their camp at Waterlooville, boarding their ship on the following day. For those on board the key objective was to keep the men occupied so that they would not dwell too much on the postponement. Thus, the 2nd East Yorks' medical officer carried out a final foot inspection, while the padre collected all letters and addresses which might provide the enemy with intelligence. The 2nd Devons had a day of recreation organized by one of the BBC's correspondents, Howard Marshall. The 6th Durham Light Infantry, on the other hand:

'During the day, two craft loads in the morning and two in the afternoon visited the Transit Camp set up by Movement Control in the sheds on the quayside. Men had a chance to have a good wash and were served with a hot meal. Camp provided writing facilities, enter-tainments, NAAFI and organised games and gave all ranks a chance to stretch their legs after the rather cramped conditions aboard the landing craft.'

On 5 June the various naval forces set sail, steaming to an assembly area off the Isle of Wight and then, having formed columns, changing course to the south-west towards Normandy. The 2nd Royal Ulster Rifles War Diary noted that 'instead of expected high tension in the face of such a mighty undertaking the feeling appeared to be calm, as if yet another of the many exercises on similar lines was about to take place'. The Channel sea was still rough and many were seasick.

During the early hours of the morning of 6 June the airborne operation was launched, with the US 82nd and 101st being dropped in the west and the British 6th Airborne Division in the east. The plan was for its airlanding brigade to come in on the evening of D-Day, with one small but vital exception. Major John Howard and six platoons of the 2nd Ox and Bucks had been given the task of seizing two vital bridges, one over the Caen Canal, and the other over the River Orne. Remarkably, his own glider landed within 50 yards of the former and the other two with him only a little further away. The Germans guarding the bridge were taken totally by surprise and it was seized intact with only a few casualties. As for the bridge over the Orne, the three gliders here landed at some distance from their objective, but when the Ox and Bucks arrived at the bridge, they discovered that the Germans had fled. Thus ended a short but highly spectacular and successful operation.

The amphibious assault force had meanwhile ploughed its way onwards through the night. The Royal Ulster Rifles commented in their War Diary that 'the huge convoy of which the Battalion was but a part, and the enormous number of Allied aircraft seen making for the Continent kept spirits buoyant'. The infantry in their Landing Ships Infantry (LSIs) had first to transship to their Landing Craft Assault (LCAs) prior to making the final run into the shore. This took place at what was called the Lowering Point, which was some six miles offshore. The 3rd Infantry Division's assault force was organized into groups as follows:

GROUP	COMPOSITION	LEAVE LOWERING POINT	TOUCH DOWN ON BEACH
		(time in minutes before (−) and after (+) H-Hour)	
1	Duplex Drive (DD) Shermans of 13th/18th Hussars	H−125	H−7.5
2	Two coys each of 2nd East Yorks and 1st South Lancs, 5 Asslt Regt RE (with Armd Vehs Royal Engineers) with 7 troops of 22nd Dragoons (flail tanks), Royal Marines Armd Support Regt (Centaur tanks)	H−90	H−Hour (0725)
3	8 LCT(R) (rocket firing Landing Craft Tank)	H−76	–
4	33 and 76 Fd Regts RA	H−65	H+75, H+105, H+195
5	Follow-up assault coys 2nd East Yorks, 1st South Lancs	H−60	H+20
6	No 4 Commando Advanced HQ, 1 Special Service Brigade, 263 Fd Coy RE	H−50	H+30
7	8 Inf Bde vehicles, one sqn 13th/18th Hussars	H−40	H+45
8	1st Suffolks	H−20	H+60

Thereafter came the remainder of 1 Special Service Brigade and further vehicles and stores for 8 Brigade, followed by 185 Brigade and then 9 Brigade, whose vehicles represented the last assault group and would land at H+360.

As 3rd Division's first infantry wave drew away from the Force Head-quarters ship, HMS *Largs*, a bugler of the 2nd East Yorks sounded the General Salute, which was acknowledged by the Divisional Commander, Major General Rennie, and Naval Force Commander, Admiral Talbot. In another LCA Major 'Banger' King of the same battalion broadcast extracts of Shakespeare's *Henry V* over the tannoy. As they closed to the shore Major Rouse of the South Lancashires:

'The noise was so continuous that it seemed almost like a siren. The seamanship was magnificent. The LCAs weaved in and out of the obstacles and we almost had a dry landing. I have very little recollection of wading ashore, there was too much going on above and around to notice it. It was, however, apparent from the beginning that it was by no means an unopposed landing. Mortar fire was coming down on the sands, an 88mm gun was firing along the line of the beach and there was continuous machine-gun and rifle fire. Immediately ahead of us a DD tank, its rear end enveloped in flames, unable to get off the beach, continued to fire its guns.'

An early casualty was the Commanding Officer of the South Lancashires, Lieutenant-Colonel Richard Burbury, who had gone ashore carrying a flag in regimental colours to act as a rallying point, but proved too easy a sniper's target. Two of their company commanders were also killed before the beach defences were overcome. On 50th Northumbrian Division's beach the 1st Hampshires landed in two feet of water, but initially received little fire. It was only when they started to tackle the dunes beyond the beach that they became involved in a tough fight, especially since little of their supporting armour had managed to make it to the beach. This was in contrast to their sister battalion in 231 Brigade, 1st Dorsets, who were able to begin moving inland within an hour. The Northumbrians' own integral 69 Brigade of Wadi Akarit Thug memory also made good progress, with CSM Stanley Hollis of the 6th Green Howards successfully tackling two pillboxes on his own. For this and subsequent actions, which served to inspire his battalion, he was awarded the Victoria Cross, the only one won on D-Day. Suffice to say that by the end of the day 50th Northumbrian Division had penetrated to a depth of six miles, and 3rd Division about four.

The battle for Normandy was, however, to be no walkover and there were to be six gruelling weeks of fighting before the Allies were able to make the decisive break-out. The Germans quickly recovered from their initial surprise and conducted a dogged and, at times, fanatical defence, in spite of overwhelming Allied air supremacy and the effects of naval gunfire,

8. British infantry tackle an assault course at Aldershot during the
middle period of the Second World War. The second man from the
left has a Thompson machine gun, which was succeeded by the Sten gun.

9. A piper entertains men of the reconstituted 51st Highland Division,
together with some Australians, in Egypt, late Summer, 1942. Each
company in a Highland battalion had its own piper.

10. British infantry advance to contact in open order, Tunisia, spring 1943.

11. Grenadiers reconnaissance platoon on patrol in Tunisia.

12. British infantry disembark from an American-crewed Landing Ship Tank (LST), Salerno, September, 1943.

13. Passing a knocked out German 75mm anti-tank gun during the break-out from the Gustav Line, Italy, May, 1944.

14. A patrol in the Arakan, Burma, 1944.

which made a deeper impression on them than even the attacker's total dominance of the air.

One major factor in the German favour was the nature of the terrain. The Normandy *bocage* is characterized by small fields bounded by high earth banks topped by hedges. The feeling of claustrophobia was increased by the ripening crops. The Germans fought for every copse, every village and every knoll and in between times there were the thumps of artillery, the crump of mortars, staccato bursts of machine-gun fire and the single crack-thump of the sniper's rifle. Casualties were heavy; by 28 June 3rd Division's had exceeded, 3,500 since D-Day. At times the strain became intolerable. A new commanding officer took over the 6th Duke of Wellington's Regiment in 49th Division on the evening of 27 June after the previous colonel had become a casualty. Three days later he wrote a report on the state of the battalion. He pointed out that in the past two weeks no less than twenty-three officers and 350 other ranks had become casualties. In the past three days he had lost two battalion seconds-in-command and a company commander, and all twelve of the original officers left were junior. Most of the men were 'jumpy' under shellfire and there had been five cases of self-inflicted wounds in three days. Officers and NCOs no longer wore badges of rank, which made life difficult since there had been so many new replacements that few men knew one another. Finally, 'I have twice had to stand at the end of a track and draw my revolver on retreating men.' He stated that the Battalion was not fit to take its place in the line and that it should either be sent back to Britain to refit or be disbanded. Montgomery chose the latter course and sacked the Commanding Officer for 'displaying a defeatist mentality' and not being 'a proper chap'.

Part of the problem was that training in Britain had not been sufficiently geared to coping with the singular tactical problems posed by the *bocage*. Consequently, the troops were mentally unprepared for it and when they automatically applied their battle drills they were disconcerted when they did not work. As a German report put it: 'The British infantryman is distinguished more by physical endurance than by special bravery. The impetuous attack, executed with dash, is foreign to him.' In particular, there was little attempt at infiltration tactics and a marked hesitancy to secure objectives quickly and exploit success, thus making the infantry vulnerable to counter-attack. The deployment of individual reinforcements to battalions engaged in heavy fighting as opposed to being at rest also caused difficulties. John Horsfall of the 1st Royal Irish Fusiliers commented on this in the context of Tunisia: 'I know of no worse ordeal for men than to join their regiments, for the first time, in the line. They did not come with their friends as we all did, nor could they know how comradeship makes up for most circumstances, however ghastly. Of all fears, the unknown is usually the worst, but these men would be lonely and homesick too. Then their reception – mud, and a wet black night, and the vicious scream of a shell.' A few weeks later, during a stiff action three of his men had fled. 'New boys of course and they couldn't know,' he commented laconically.

Yet morale, and hence fighting efficiency, was maintained among the vast majority of battalions in Normandy. An officer of the 1st Suffolks listed three points in his battalion which helped to achieve this:

'Our Padre (the Reverend Hugh Woodall) allowed no dead bodies to remain without burial, highly dangerous work since he personally saw to collection as well as burial. The Medical Officer (Captain Robinson, later killed) had casualties collected by the stretcher bearers at the double, therefore all ranks had confidence in his casualty collecting procedure. The Quartermaster got the rations up somehow, usually on carriers, and always at the time ordered, for as one of my five COs said "Feed them and you can do anything with them".'

Over the particular problem of recovering wounded in standing crops General Hubert Essame, who commanded an infantry brigade in Normandy, noted:

'A soldier who is hit when advancing through corn of this height $(2\frac{1}{2}$ ft) will fall to the ground and be invisible to the stretcher-bearers looking for him. His comrades know this and have to go on with the advance. They are also well aware that a wounded man may die if left unattended for any length of time . . . The sights and sounds of the wounded can create tensions too great for soldiers unused to battle . . . When a man was hit a comrade stopped briefly, took his rifle and bayonet and stuck it in the ground beside him and placed his steel helmet on top of the rifle to guide the stretcher-bearers . . . it was a poignant sight to gaze on these rifles surmounted by their tin helmets looking like strange fungi sprouting up haphazardly throughout the cornfields.'

Yet, as the Germans rightly identified, it was the infantryman's traditional doggedness which won through in the end. An RAF wing commander in Normandy:

'I think that one of the things I shall never forget is the sight of the British infantry, plodding steadily up those dusty French roads towards the front, single file, hands bent down against the heavy weight of all the kit piled on their backs, armed to the teeth; they were plodding on, slowly and doggedly towards the front with the sweat running down their faces and their enamel drinking-mugs dangling at their hips; never looking back and hardly ever looking to the side – just straight in front and down a little on the roughness of the road; while the jeeps and the lorries and the tanks and all the other traffic went crowding by, smothering them in great billows and clouds of dust which they never even deigned to notice. That was a sight that somehow caught at your heart.'

Eventually, on 25 July, the Americans began their break-out from St Lô and the British Second Army also began to emerge from the *bocage*, initially in an operation, BLUECOAT, on the American left flank and designed to keep the German armour away from St Lô. The star role was taken by 11th Armoured Division, whose task it was to screen the right flank of the British attack, but which found itself breaking through the German lines in spectacular fashion. General Pip Roberts, the Divisional Commander, had led it in GOODWOOD, the abortive attack by three armoured divisions west of Caen two weeks earlier and realized that a weakness in the organization of an armoured division operating in relatively close country was a lack of infantry close up with the armour in order to tackle strongpoints, especially on the flanks.

Consequently, Roberts reorganized his two brigades so that each had two armoured regiments and two infantry battalions. To find the fourth armoured regiment the divisional reconnaissance regiment, equipped with Cromwell tanks, had to largely forsake its normal role. Furthermore, while both armour and infantry were by now used to working on a squadron/company basis, he was told by his Corps Commander, General Dick O'Connor, that because they would be operating in extreme *bocage* it was essential that this co-operation be at troop/platoon level. Roberts also stressed to his brigadiers that this new organization was totally flexible and that they must be prepared to command any combination of armour and infantry. Roberts used this system for the rest of the campaign and found it 'highly satisfactory'.

Even so, even the motor battalion's carriers were not as well protected as the tanks, and this would always act as a brake. One system used to overcome it was for the infantry to ride on the tanks, but, while this was acceptable during an approach march, once the tanks came under fire the last thing they wanted was infantrymen interfering with the crews' vision and the turret traverse, and they would find themselves unceremoniously dumped on the ground. The solution was found by the Canadians, who removed the turrets from Sherman tanks and converted them into armoured personnel carriers (APCs). Known as Kangaroo Rams, each was capable of carrying eight infantrymen and by the end of 1944 two regiments' worth, each able to lift a battalion, were available and controlled by 79th Armoured Division.

As for BLUECOAT itself, the 8th Rifle Brigade, the motor battalion in 11th Armoured Division, who were operating with the 23rd Hussars, noted at the end of the second day that they were 'now through the enemy's gun line'. As for the subsequent advance:

> 'If there were flags on the houses it was a sure sign that the enemy had gone. If we entered a village in silence, with no flags, no welcoming cheers, no children, only eyes watching silently from windows, stray dogs, a cart, perhaps overturned, then one could expect to meet the Germans round the corner.'

Unfortunately 11th Armoured Division was not able to exploit to the maximum. This was for a number of reasons. Firstly, they were operating on the inter-Allied boundary and there was confusion over exactly where this ran on the map. Indeed, at one point they became hopelessly intermingled with a US infantry regiment. Their axis was also taking them south to the communications centre of Vire, but this was an American responsibility and was also taking 11th Armoured too far away from the main thrust of the British attack, which was to the south-east. Finally, they had to pass through the large Fôret l'Evêque, which had few routes through it and the 3,000 vehicles of the division became very snarled up. They were ordered to swing south-east and then became embroiled in a counter-stroke by 9th SS Panzer Division from the west and also came under threat from 3rd Parachute Division. There followed a desperate battle around the Perrier Ridge. A soldier in the 3rd Monmouths:

> 'For two days the cooks' trucks failed to get through. We had been chewing away at the hard biscuits we always carried. We couldn't leave them alone. Some of the worst addicts had eaten them all. We all carried an emergency tin of chocolate but we dared not touch that.
>
> 'But worst of all was the feeling of tiredness; for days we had no more than two consecutive hours of sleep. The drain of casualties had made the burden of guard duty and patrols heavy for those who remained. We just couldn't catch up with lost sleep. Dusty, begrimed, sore-eyed, we would doze off at odd times only to be wrenched back to sweaty consciousness by the "woof" and crump of another bombardment.'

Later they were relieved by the 1st Royal Norfolks of 3rd Division, but in the midst of the relief they were suddenly attacked by 10th SS Panzer Division. The 'Normons', as they temporarily called themselves, repulsed it, but the Norfolks in particular suffered heavily, having been caught in the open by the opening barrage. However, Corporal Sidney Bates, commanding a forward section which was in danger of being cut off, charged a group of 50–60 Germans, firing a Bren from the hip. Three times he was wounded, the last fatally, but his single-handed action averted a minor catastrophe and won him a posthumour Victoria Cross. Eventually the Germans withdrew and the battle died down.

The final key action of the Normandy fighting was the closing of the Falaise pocket. After this the British and Canadians broke out over the Seine and, in the mad dash that followed, crossed the Albert Canal and were approaching the Dutch border before the ever-stretching supply lines finally snapped and they ground to a halt, out of fuel.

The next major event was MARKET-GARDEN, Montgomery's ambitious plan to turn the German northern flank by seizing bridges over the main waterways in Holland. While the US 82nd and 101st Airborne Divisions were successful, the British 1st Airborne Division, tasked with securing the bridge over the Rhine at Arnhem, failed after an epic battle. Their airlanding brigade consisted of the 2nd South Staffords, whose Major Robert Cain and Sergeant John Baskeyfield both won Victoria Crosses, 7th King's Own Scottish Borderers, and the 1st Border Regiment, was initially responsible for guarding the dropping zone, but then became embroiled in the fighting in Arnhem itself.

On the ground XXX Corps, which was to advance and link up with the airborne forces, had a frustrating time, its axis being restricted to little more than a single road for much of the way because of the marshiness of the terrain. Initially the Guards Armoured Division took the lead, but later handed over the baton to 43rd Wessex Division as the terrain became less and less suited to armour. Indeed, so bad was it in terms of high banks and dykes that the 7th Somerset Light Infantry had to attack the village of Oosterhoot, which was held by a battalion supported by tanks and self-propelled guns, with no armoured and virtually no artillery support. They made two valiant attacks and suffered heavily, before it was eventually taken at the third attempt. This put them just ten miles from the Lower Rhine and at last light an armoured group consisting of the 5th Duke of Cornwall's Light Infantry, a squadron of the 4th/7th Dragoon Guards, and a machine gun platoon of the 8th Middlesex set off on the final dash to the river. They succeeded in linking up with the Polish Parachute Brigade, but in the meantime German troops, including five Tigers, managed to infiltrate the column. A tank-hunting party had to be sent back to tidy up the situation and in the course of this Private Brown of the 5th DCLI took on a Tiger at pointblank range with his PIAT, destroying it, but was blinded by the blast. As he was carried away, he said: 'I don't care, I knocked the —— out.'

The 4th Dorsets were now ordered to cross the river in assault boats in order to establish a physical link with the Paratroops at Arnhem. This they attempted on the following night, 24th/25th, but lost most of their boats and some 250 men. This put the seal on MARKET-GARDEN and 1st Airborne Division was ordered to withdraw across the river. Less than a quarter of its 10,000 men succeeded in doing this; the rest were dead or made prisoner. The failure of the operation meant that the prospect of the war in Europe being over by the end of 1944 vanished. The South Staffords and the Border Regiment subsequently adopted a flash of a yellow glider on a green background to commemorate Arnhem.

As early as July, 1944, General Sir Ronald Adam, the Adjutant-General, had visited Normandy in order to warn Montgomery and British Second Army Commander Sir Miles Dempsey that the manpower situation was becoming critical in view of the large number of infantry casualties. If they continued at current rates it would be impossible to replace them and divisions would have to be cannibalized. This was the reason 6th Airborne Division and the Commandos were retained for so long in Normandy, the original plan being to withdraw them to England after the first few weeks. A sign of how desperate the situation was becoming was that some 7,000 RAF conscripts had to be hastily transferred to the Army. By October the position had worsened and Montgomery was forced to disband 59th Division. At the same time a number of divisions in England were also disbanded. These were 38th, 45th, 47th, 76th and 77th, together with the 80th Division, which had only been formed the previous year. Finally, towards the end of 1944, Montgomery was forced to break up 50th Northumbrian, and this in spite of a personal plea by Churchill for its retention.

To overcome the shortage of junior infantry officers, help had been given by the Canadian Army. This was a result of two divisions in Canada being disbanded, thus throwing up a surplus of officers. As a result the Canadian Government offered their services to the British Army. This was accepted and in Spring, 1944, the officers, over 600 of whom volunteered, were interviewed by a selection board and sent for a short refresher course in Canada prior to being posted to British battalions. In 3rd Infantry Division alone, 39 CANLOAN officers served, of whom ten were killed in action and six decorated for bravery. A more unconventional way of overcoming a temporary manpower shortage was adopted by the 1st Rifle Brigade, who incorporated a number of young French and Belgian volunteers in their ranks.

One major amphibious operation was mounted before 1944 was out and this was against the island of Walcheren in the Scheldt estuary so that the port of Antwerp could be opened up and hence shorten the Allied supply lines. Before this could happen the Canadians had to clear the Germans out of the Breskens pocket on the mainland south of the island. To do this they were reinforced by 52nd Lowland Division, which was now finally to see action. The Division had relinquished its mountain warfare role in the summer and had begun to convert to an airlanding formation. The Divisional Commander had offered one brigade to reinforce 1st Airborne Division at Arnhem, but General 'Boy' Browning, commanding I Allied Airborne Corps, had turned this down. Now it was finally to be given its chance, but as a conventional infantry division. This time it would enjoy more success than the last time British infantry had fought on Walcheren, in 1809. For the Breskens pocket 157 Brigade was earmarked to support the Canadians. This overcome, the 4th/5th Royal Scots and 6th Cameronians of 156 Brigade crossed from the mainland to land on South Beveland, the neighbouring island to Walcheren. This involved a nine-mile voyage in Buffaloes, American tracked amphibians, and the landings were successfully achieved. Later, 157 Brigade was also brought across to South Beveland.

Finally, to tackle Walcheren itself, Commandos landed at Flushing, with 155 Brigade as the follow-up, while a further Commando force landed at Westkapelle on the western nose of the island. The final attack, which sealed the fate of the German garrison, was a surprise thrust on Middelburg, in the centre of the island, by the 7th/9th Royal Scots, once more putting the Buffaloes to good use in overcoming the floods. Before November was out and, after an extensive naval operation to clear the estuary of mines, Antwerp was once more open to shipping.

There was now a relative lull in offensive operations, at least as far as 21st Army Group was concerned, although to their south the Americans were heavily involved in closing up to the West Wall, at least until they were suddenly struck by a German counter-offensive in mid-December. One battalion to take advantage of the lull was the 1st Suffolks. During the withdrawal to Dunkirk in May, 1940, they had been forced to leave their drums in the safekeeping of French civilians at Bouraix. Lieutenant-Colonel Dick Goodwin took the opportunity to send a party back to the town and the drums, which had been hidden in hat boxes in a factory, were recovered. The Battalion held a special parade in Holland to welcome them back and three members of the 1940 Corps of Drums were there to beat them once more. Other battalions had also hidden theirs, having taken them to France to act as a symbol in lieu of the Colours, but were not so lucky. The 2nd Worcesters buried theirs, but eventually got only two back, one from Denmark.

As part of its Christmas Day, 1944, broadcasts, the BBC decided to interview a soldier of the British Liberation Army and get him to speak of his hopes once the fighting had ended. It was to be run immediately before the King's Christmas Message. It was appropriate that an infantryman should be selected for this, Corporal Bob Pass of the 1st/5th Queens in 7th Armoured Division, and a veteran of Dunkirk, Crete, North Africa, Sicily and Italy, and once a window cleaner in Brixton, south-east London. It was made even more symbolic by the fact that The Queens were holding the village of Wehr, just inside Germany. Their two reserve companies were, however, able to hold a service in the village Gasthof and to have a proper Christmas meal afterwards. That night, from across the frost-bound no-man's-land floated the singing of *Heilige Nacht*. Corporal Pass's broadcast was entitled 'The Journey Back', but sadly he never made it. Just over two weeks before the end of the war in Europe he and his patrol were flushing some Germans out of a wood. One raised a white flag and Pass went forward to meet him, only to be fatally shot by the others.

Once the Ardennes counter-attack had been absorbed and the situation restored, the Allies now began the operation to close up to the Rhine. After a preliminary operation called BLACKCOCK, the main British-Canadian offensive, VERITABLE, was launched on 8 February. Initially it went well, thanks to a massive artillery bombardment, larger than any fired by the British Army during the war, and to the fact that the German units manning the first line of defences were of low medical grade. Flooding of much of the area by the Germans, combined with heavy rain, began to slow progress on Day 2, as did the deployment of the seasoned fighters of Schlemm's Third Parachute Army. The brunt was borne by II Canadian Corps and Brian Horrocks's XXX Corps, the British infantry divisions involved being the 15th Scottish, 43rd Wessex, 51st Highland and 53rd Welsh. The battle was to last five long weeks. Eisenhower commented: 'Probably no assault like this in war has been conducted in more appalling conditions of terrain than this one.' The 51st Highland and 53rd Welsh found themselves battling through the sinister woods of the Reichswald. Horrocks himself described what conditions were like for these two divisions:

'Their forward troops would very often consist of two young men, crouching together in a fox-hole, both of whom had long since come to the conclusion that the glories of war had been much over-written. They were quite alone for they might not be able to see even the other members of their own section and all around them was the menace of hidden mines.

'It is this sinister emptiness that depresses them most – no living thing in sight. During training, officers and NCOs had been running round the whole time, but they cannot do it now to anything like the same extent, or they won't live long. Our two young men are almost certainly cold, miserable and hungry, but they are at least reasonably safe as long as they remain in their fox-hole. But they know that soon they will have to emerge into the open to attack. Then the seemingly empty battlefield will erupt into sudden and violent life. When that moment arrives they must force themselves forward with a sickening feeling in the pit of their stomachs, fighting an almost uncontrollable urge to fling themselves down as close to the earth as they can get. Even then they are still alone amidst all the fury, carrying their loneliness with them.'

It was at times like these that the strengths of the regimental system, with its capbadge loyalties and bonds developed among men with common backgrounds and interests, came into its own.

For the 15th Scottish and 43rd Wessex Divisions, on the other hand, VERITABLE was marked by costly street fighting in the battles for such towns as Cleve, Udem and Goch. Something of what it was like can be gleaned from the War Diary of the 1st Gordons during the capture of Goch:

'At this time it was sufficiently light to see a man 50 yards away but too dark to distinguish friend from foe. The enemy had been roused by bursts of sten fired by 8 Pl inside the house which they were clearing. 7 Pl had been ordered to clear the main farm building. What happened to them is not fully known. They went towards it at the same time as 9 Pl and Coy HQ reached their buildings, and a straggler reported afterwards that as they approached the building they saw one or two men in front of it. Thinking that it might be our own men they shouted to them and the reply was a burst of fire. Lieut C. C. Hewitt's body was found next day and it would appear that the Pl became scattered after he was killed. Meanwhile Coy HQ was firm in their building and killed one or two Germans who tried to run into it under the impression that it was empty. The Bn Comd then visited 9 Pl in the next building and found that Sgt CLEVELAND's Pl had been unable to clear the buildings beyond it as any movement outside to the back was met by spandau fire from trenches 100 yards beyond. The Bn Comd formed the opinion that . . . more men would be required to clear the remaining buildings . . . Meanwhile 8 Pl had not been able to clear the second house and the Bn Comd found Cpl HENDERSON firing a PIAT at it from just in front of the first one.

He was told to put smoke down and rush the house and then, having cleared it, take the Pl forward to Capt KYLE. Full details of what happened after this we shall not know until Capt KYLE and the NCOs that were taken prisoner with him return. At the time of writing we only know that he and 46 ORs are missing and that several casualties were later recovered from the battlefield.'

Goch cost the Gordons three officers and twenty-one men killed, seven and fifty-nine wounded, and one and forty-eight missing, and illustrates just how difficult this form of combat is. What is especially significant are the casualties among junior officers, with subalterns accounting for all but two of the officer casualties. Indeed, and illustrative of how tough the fighting was in North-West Europe for the infantry, no less than 102 officers served with the 1st Gordons between 6 June, 1944, and 27 March, 1945. Of these fifty-five were rifle platoon commanders, whose aggregate casualties by the time the Battalion crossed the Rhine were fourteen killed, thirty wounded and eight invalided; their average length of service with the Battalion worked out at a mere thirty-eight days. To set this in context, the infantry subaltern in France during 1914–18 spent an average of six months with his battalion before becoming a casualty of being posted.*

Eventually the Germans were driven back across the Rhine and the Allies set about preparing for the crossing of the last main obstacle which barred the way into the German heartland. Montgomery planned five simultaneous night crossings. These would be spearheaded by 51st Highland at Rees, 15th Scottish at Xanten, 1 Commando Brigade at Wesel, and two American divisions north and east of Rheinberg. XVIII Allied Airborne Corps (6th British and 17th US Airborne Divisions) would drop and seize the high ground east of the Wesel crossing during the following day. Careful preparation ensured success, but, as often happens in battle, even the best laid plans can go wrong. Wynford Vaughan-Thomas of the BBC crossed in a Buffalo with 15th Scottish Division:

'We were across, untouched. Well there was an immense feeling of relief and excitement throughout our little party. The Commanding Officer gave a signal, the piper lifted his pipes to his lips, and he blew, and only an agonised wailing came from his instrument. Again he tried, and again the wail. If ever a man was near to tears, it was our piper. His great moment, and now, as he cried in despair: "Ma pipes man, they'll no' play".'

* The often quoted claim that a subaltern lasted a mere three weeks, especially during the Somme and Third Ypres battles, has been totally disproved through analysis of battalion War Diaries.

The Commandos, thanks partly to the airborne operation, experienced little difficulty at Wesel, but both 15th And 51st Divisions had problems from German counter-attacks. Indeed, the popular commander of the latter, General Tom Rennie, who had previously been wounded by a mine while commanding 3rd Division in Normandy, was killed by a mortar bomb, and 15th Scottish had to face a furious counter-attack by 15th Panzer Division. The 1st Black Watch bore the brunt of this and the fighting became particularly severe and confused around the village of Speldorp. Nineteen-year-old Lieutenant John Henderson volunteered to take out a patrol to try and clarify the situation, but soon came under heavy fire. Undaunted, he continued forward with just a Bren gunner, ordering the remainder of his men behind cover. A burst of machine-gun fire killed the gunner and knocked the revolver out of Henderson's hand, but he immediately rushed the post, killing the crew with a shovel. He then occupied a building which was on fire with his patrol, crawling back himself under heavy fire to rescue the Bren from the dead gunner. Thereafter he and his men held their house against repeated attacks until relieved twelve hours later by Canadians. Henderson's actions were rewarded with a DSO, this in itself an unusual reward for a subaltern, although they were well worth a Victoria Cross.* Eventually the bridgeheads were secured and 21st Army Group could, to use Montgomery's words, 'beat about the North German Plain'.

The last weeks of the campaign were, however, by no means easy. There was still plenty of fight left in some groups of Germans, usually fanatical Nazis whose only wish was to die for the Führer and to ensure that before they did so they took as many of their enemies as possible with them. The troops also had to endure the horrors of liberating concentration camps, such as Bergen-Belsen, and the happier task of freeing prisoners of war, many of whom had been 'in the bag' since 1940. At one of these, Stalag XIB at Fallingbostel, the 8th Hussars, who liberated it, were greeted by an immaculately turned out guard, complete with blancoed webbing. So smart was it that the Hussars thought that 6th Airborne Division had beaten them to it. The man behind this was John Lord, a Grenadier, who had been wounded and captured at Arnhem while RSM of the 3rd Parachute Battalion. The camp consisted of groups from almost every nation in Europe and a large number of Americans. The British compound had both men from Arnhem, many of whom were wounded, and other long-term POWs who had been marched from the east in the face of the Russian advance. Most were in a very poor physical and mental state, not helped by the appalling way in which the camp was administered and the starvation

* Some 24 years later John Henderson conducted the service at my own marriage.

127

rations. Lord, as the senior warrant officer, took charge and ran it on Guards Depot lines. Soon the men recovered their self-respect and, in many cases, their health. Later, 'Jackie' Lord, as he became affectionately known, was to leave his indelible print on generations of officer cadets, including the author, at Sandhurst, where he was Academy Sergeant Major for many years.

Late on 4 May HQ 21st Army Group transmitted a signal to its subordinate formations: 'Cancel all offensive ops forthwith and CEASE FIRE 0800B hrs 5 May 45.' Major (later Major-General) Glyn Gilbert, a company commander in the 2nd Lincolns, reflecting on the fall of Bremen, 3rd Division's last operation:

'"That's about it then," said my old friend CSM Sam Smalley. He was, as usual, right. We had fired our last shots in Europe. There were five of us left in C Company who had landed in Normandy in June last year, and I remember suddenly feeling very tired and sad about the casualties we had suffered during the past few hours. I did not realize it at the time, but I was the only rifle company commander in the Division [3rd] who had landed on D-Day and not been killed or wounded.'

As after Waterloo, however, it was to be a little while before the infantryman with his rifle and bayonet entered the enemy's capital and therefore put the seal on the victory. It was on 4 July, after negotiations with the Russians, that 7th Armoured Division finally entered Berlin, with the infantry being represented by the 1st Grenadiers, 1st/5th Queens and 2nd Devons. It was the end of a long road.

CHAPTER NINE

Forgotten Armies
Italy and Burma (1944–45)

THE Allied forces in Italy were conscious from the beginning of 1944 that the public spotlight had switched from them to the opening of the Second Front. This was accentuated after the Normandy landings, by which time they had realized that their main task was merely to keep as many Germans as possible tied down in Italy. Indeed, the British troops ruefully called themselves 'The D-Day Dodgers'. In terms of the intensity of the fighting and the often inhospitable climate and terrain, though, Italy was no mere sideshow.

The main problem facing the Allies at the beginning of the year was how to penetrate the Gustav Line. The solution they came up with was frontal attacks against the German defences combined with an amphibious landing at Anzio to the rear. This, it was hoped, would unhinge the Gustav Line and trap a significant portion of the forces holding the western part of it. The landings themselves, conducted by 1st British and 3rd US Divisions, under the control of VI US Corps, were mounted on 22 January and initially took the Germans by surprise.

Unfortunately, General Lucas, the Corps Commander, partially because of ambiguities in his orders, failed to take advantage of this. Instead of breaking quickly out of the beachhead, he allowed the Germans to concentrate forces and to begin counter-attacking in force. The experience of the 2nd North Staffords is typical:

'On the 7th of Feb – the quietest day since landing – something was certainly brewing, no shelling or mortaring and hardly a sound from either side until 10 o'clock at night when the forward coys reported enemy infiltrating through the platoon positions. At first it was thought that this was merely patrol activity but at 2300 hours the intensity of artillery, mortars and small arms fire were indicative of an attempt to break through. The forward coys – A, B and C – were almost surrounded within an hour of the commencement of the attack, and grim hand-to-hand fighting in slit trenches and thick undergrowth raged until two in the morning when it was obvious that the attack was developing into a major thrust. Enemy tanks and self-propelled artillery closed in and poured shells into coy localities and supply routes, and hordes of yelling fanatical infantry swarmed onto the main BUON RIPOSO ridge. A and B Coys, having sustained terrific casualties, were regrouped and to prevent further infiltration took up positions immediately in front of Bn HQ.

'Meanwhile C Coy held their original position, and, although completely surrounded, contested the issue until completely over-powered by sheer weight of numbers.

'At approximately two in the morning there was a slight lull, the enemy obviously had to regroup his assault troops who, judging from the amount of shrieks and groans of the wounded had paid an enormous price for the ground gained.

'All this time the Divisional artillery had never ceased firing – and the number of shells hurled at Kesselring's men was phenomenal; many German prisoners testify to the accuracy and devastating effects of our shelling. At two thirty, however, the attack was renewed with unbounded ferocity and in spite of all efforts the Bn was forced to yield some ground; sheer weight of numbers, in addition to tanks and SP guns, decided the day.

'On the right the Grenadier Guards [5th Battalion] were in the same predicament and towards first light when it was known that 3 Inf Bde were coming to counter-attack the remnants of the two battalions were withdrawn.

'With this withdrawal brings a story that will live for ever in the minds of those who were present. A concentration area on the edge of a wood was chosen for the Bn rallying point. Including B Echelon personnel, 230 was all that could be mustered and, as the roll call was completed, a concentration of shells came down on the Bn area causing a further 26 casualties, six of which were fatal.'

In all the North Staffords suffered twenty-three officers and 300 men killed, wounded and missing during the two days' fighting. In less than ten days, having received reinforcements, the Battalion was occupying depth positions in the beachhead. Two days after it had done so, it was called upon to make a counter-attack to restore a critical situation. This it did, carrying all before it, literally at the point of the bayonet, thus providing as good an illustration as any of how quickly a battalion could recover from crippling losses.

The fighting to hold the Anzio beachhead was to continue into the Spring, but elsewhere there was just as, if not more, formidable a battle. This was to seize the prominent Monte Cassino, a key point in the Gustav Line and crowned by a famous monastery of that name. During January–May, 1944, no less than four major assaults were made against this feature, one that came to symbolize the fighting in Italy more than any other. It was not just that it was defended by German Paratroopers, probably, at least in the eyes of the British infantryman, the most respected element in the German armed forces, but also that so many nationalities took part in the attacks against it. Americans, Free French, British, Indians, New Zealanders and Poles, who eventually captured it, all suffered heavily.

One example among many of how tough the fighting was concerns the 1st Royal Sussex in 4th Indian Division. They were tasked with seizing Point 593, which commanded the approach to Monastery Hill. They tried first with a company attack on the night of 15/16 February. Making a silent approach, which was very difficult in view of the large number of loose stones and difficult ground, they advanced only 50 yards before they came under withering machine-gun fire and a flurry of grenades. Groups tried to get round the flanks, but the approaches were too steep, and eventually the survivors were forced to withdraw. Out of three officers and sixty-three men who took part, two and thirty-two had become casualties. The Royal Sussex were ordered to make another attempt the following night. No preliminary barrage was possible because the objective lay only 70 yards from the Battalion's forward positions. All the guns could do was to put down suppressive fire on the features beyond the objective. Mortars could be used, but the Battalion had few bombs and was forced to scrounge American ones, which could only be fired from captured Axis mortars brought by the Battalion from North Africa as souvenirs. The commanding officer also called for large stocks of grenades, which had to be brought up by mule. These were late in arriving because the mules had met shell fire en route, which resulted in the loss of half the grenades. The attack began an hour late and, when it did, some of the artillery rounds fell short, inflicting casualties on the forward companies. As if this was not enough, the attackers found themselves faced by unexpected deep crevices in the rocks. Even so, one party did manage to reach the objective, but as it moved beyond to consolidate it ran into German reinforcements. A desperate grenade battle erupted, but the Germans were better supplied and eventually forced the Royal Sussex back to their own lines. They had suffered another ten officers and 130 men killed, wounded or captured.

The following night it was the turn of the 4th/6th Rajputana Rifles, this time supported by the Royal Sussex, but again the attack failed, the Rajputs losing nearly 200 men. This marked the end of the second battle for Cassino, but it meant no relief for the Royal Sussex, who had to endure a further month in their exposed position.

Eventually, at the end of May, the Allies finally managed to break through the Gustav Line, link up with the troops in the Anzio beachhead and drive on to Rome, which was entered by Americans on 5 June. The Irishmen of 38 Brigade marked this in a very special way, with a formal visit to the Pope in the Vatican by 150 men from the 1st Royal Inniskilling Fusiliers, 1st Royal Irish Fusiliers and 2nd London Irish Rifles, accompanied by their Pipes and Drums, who played *The Minstrel Boy* as the Pope received them. A special Mass in St Peter's followed, after which the Pipes and Drums beat Retreat outside.

The drive north did not cease, however, even though it was now weakened by the loss of American and French troops, who left the theatre to take part in the landings of the south of France, which were mounted in August. Kesselring had prepared a new and just as formidable defensive position as the Gustav Line. This was the Gothic Line, which took maximum advantage of the Apennines. At the beginning of September the Canadians and Poles, operating on the coastal plain of the Adriatic, managed to break through it, but the Eight Army was then faced with a seemingly endless succession of river lines to overcome. For the US Fifth Army in the west the situation was made even more difficult by firstly having to break through the Arno Line, which lay some twenty miles south of the Gothic Line, and by the fact that the Appenines ran laterally across their front. Consequently, the Americans became bogged down in the mountains amid the autumn rains.

One British division, the now veteran 78th, which took part, had a particularly tough time. The Irish Brigade captured Point 382 only on its third attempt and Monte Spaduro at the fourth. Monte la Pieve was only secured after five attacks by 11 Brigade had failed, and 36 Brigade had to make two attacks on Monte Acqua Salata to drive the Germans off. So tough did the Division find this autumn fighting in the mountains that its desertion rate rose alarmingly until it was averaging fifteen men per battalion per month. A number of these were Tunisian veterans, some decorated for bravery and others who had already been wounded more than once. Their reserves of courage had been exhausted.

The problem was two-fold. Firstly, the Division was fighting under American command, that of General Mark Clark. As has been previously mentioned, the Americans were less sympathetic to casualties, not suffering the same manpower crisis as the British. Hence they were prepared to drive formations more ruthlessly, and tended to view the British policy of nursing the 'old steeplechaser', to use General Dick McCreery's words, as hesitancy and worse. Secondly the Battleaxe Division, as the 78th were known from its divisional sign, had established a high combat reputation and there was a tendency to deploy it where others had failed. This, however, is not to say that the high desertion rate resulted in the Division becoming non-effective; it did not, but it was an indication that too much was being asked of it.

The two other categories of deserter were the genuine 'bad hats' and the reinforcements. The former included a number of men who had previously deserted in North Africa and had been posted to Italy under suspended sentence. Often they influenced others to desert with them. Much of the problem here lay in a mistaken policy of often posting a man identified as a ne'er-do-well to the infantry in the traditional belief that this was where the dregs of society went. There was also a rumour current in Italy during Autumn, 1944, that all sentences for desertion would be lifted as soon as the war was won. This was so rife that in December the government had to issue a public denial. Even so, this had little effect, and it was not until February, 1945, that the desertion rate began to fall away. This was partially because victory was in sight, but also because of a ruling that, in addition to the mandatory three years' penal servitude,* those convicted of desertion would have all previous service disallowed when it came to reckoning their release dates.

As for the new reinforcements, this is a problem which we have already addressed. It needed good man-management to ensure that they were integrated, but often it was nigh on impossible to do this in a front-line battalion. Matters were not helped by the fact that a number of gunners found themselves transformed into infantrymen, another indication of the manpower crisis, and sent as individual reinforcements to battalions rather than being posted in groups. One factor, however, did ensure that the incidence of desertion did not rise even higher. This was that shellshock was now widely recognized as a genuine medical condition and every effort was made to try and catch it in its early stages. Often a victim could be quickly cured by a good sleep, a hot meal and sympathetic counselling.

By the end of October the US Fifth Army had come to a grinding halt in the mountains, and even Mark Clark was forced to accept that the rate of replacements was not matching that of casualties. On the Adriatic coast the Eighth Army continued to grind northwards, but, with winter now arrived, Alexander, who had been appointed supreme commander in the Mediterranean in November, decided, at the end of December, to call a halt and rest his wearied armies until the spring.

The final offensive in Italy was launched in April, 1945. The Eighth Army struck first and drove for the Argenta Gap. Then the massive weight of Allied airpower was transferred to Fifth Army to enable it to break out

---

* For those who deserted in the Anzio beachhead this term had been raised to five years, an indication of how desperate the situation became there.

of the mountains. Both armies took part in the fall of Bologna, and then the Americans moved north-west towards the Italian lakes, while Eighth Army advanced towards Trieste, which was reached on 2 May. Here they met Tito's partisans, who were bent on seizing the port, something that would become an immediate post-war problem. Both armies entered Austria and it was appropriate that, as in November, 1918, infantry should be the first to cross the border, on 8 May, VE Day. In this case it was 61 Brigade of 6th Armoured Division. This was made up of the 1st Battalion of the 60th Rifles and the 2nd and 7th Rifle Brigade. The two Regular battalions were founder members of the 7th Armoured Divison, the original Desert Rats, and, as such, had fought throughout North Africa, while the 7th Rifle Brigade had been blooded at First Alamein, and all three battalions had fought their way up the length of Italy. They were followed by the remainder of 6th Armoured Division and 78th Division. The Italian campaign had been long, frustrating and very tough. For the infantry, in particular, it had been a severe test, but one from which they had emerged triumphant, in spite of their trials and tribulations.

In Burma the battle for the defence of India still had to be won before Fourteenth Army could go over to the offensive. The first test of Slim's 'new model army' came in the Arakan in February, 1944. The Japanese had been aware for some time that the British would inevitably re-invade Burma and decided to pre-empt this by launching an offensive of their own. This was to be into Assam, but a subsidiary operation would be mounted in the Arakan prior to this in order to divert British attention. The Arakan was held by Christison's XV Indian Corps (5th and 7th Indian Divisons). Their opponents were the Japanese 55th Division, which, on 3 February, 1944, set in train a plan to infiltrate through the left flank of XV Corps and strike it from the rear. Three days later they overran the headquarters of 7th Indian Division, although the divisional commander, Frank Messervy, managed to escape capture.* It looked as though disaster threatened, but Christison had given firm orders that troops were to stand and fight, even if the Japanese had got into their rear, and this was what now happened. The main focal point of the fighting became the so-called Admin Box, the Corps Administrative Area at Sinzweya. Hastily reinforced by the 2nd West Yorkshires, 4th/8th Gurkhas and two tank squadrons of the 25th Dragoons, the Admin Box withstood a fortnight's seige. Lieutenant-Colonel (later Brigadier) G. E. Cree of the 2nd West Yorkshires described what the Box looked like when he arrived:

* This was the second time that Messervy had had his divisional HQ overrun. In May, 1942, during the fighting at Gazala he had been captured, when commanding 7th Armoured Division, but escaped during an artillery barrage.

'It was an area of flat, open ground roughly a mile square, surrounded on all sides by hills and jungle. It contained most of the 1st line and all the 2nd line MT of the Division [5th Indian], which was operating on an animal transport basis; the MDS [Main Dressing Station] supply, ordnance, ammunition, and engineer dumps stocked up with at least a month's reserves; spare mules; provost; artillery of several descriptions, and much else besides. The area was, in consequence, somewhat congested. It was organized as a "Box", that is the units and sub-units composing it were in a rough defensive position and were supposed to be able to protect themselves against minor enemy enterprises using their own weapons and man-power. The man-power consisted mainly of Indian ancillary services. They were armed and had been basically trained in the use of their weapons. But the most that could be expected of them was purely static defence; they were untried in battle and untrained in anything except firing out of a trench.'

In one early attack the Japanese managed to get into the Box's hospital, shooting doctors and bayonetting wounded, before a counter-attack drove them out. On another occasion RSM Maloney of the West Yorks and the Battalion clerks successfully ambushed a party of fifty Japanese. One Japanese officer attempted to despatch a West Yorks sergeant with his sword, only to be bayonetted by him and another NCO. The West Yorkshires also had to resort to the bayonet in order to clear off Japanese infantry guns which had set ammunition dumps ablaze. Throughout the siege the Box was kept resupplied by air and it was this and the resilience of the defenders which eventually forced the Japanese, their supplies exhausted, to withdraw. Such had been the intensity of the struggle that the West Yorkshires alone suffered 144 killed and badly wounded, but what it had finally demonstrated was that British and Indian soldiers were more than a match for the Japanese and their logistical support was infinitely superior.

What had happened in the Arakan was to be repeated on a larger scale a month later in northern Burma and Assam. The main Japanese offensive began on the night of 7/8 March, first clashing with those veterans of the first campaign in Burma, 17th Indian Division, in the Tiddim area. After a week the Division began a withdrawal northwards along the road to Imphal. The Japanese constantly got in behind the Division and set up roadblocks to bar the withdrawal. Raymond Cooper, a company commander in the 9th Border Regiment:

'The Japs, holding the hills above, put up a road block and covered it with fire. We located them with infantry patrols, sent a Battalion on a long chukker round on the flank, deployed the guns, whistled up the RAF from Imphal, and in due time reopened the road after having killed as many of the enemy as possible.'

This was in stark contrast to the 1942 retreat, and the Division arrived in Imphal with no less than 90 per cent of its vehicles and pack animals which had left Tiddim, 120 miles away. During the three weeks' operations Cooper's company's casualties had been a mere eight wounded. In the meantime two further Japanese divisions had struck some 150 miles to the north, and one of these succeeded in cutting the road north from Imphal and isolating Kohima.

Kohima itself lay at the head of a pass on the Imphal-Dimapur road and was 5,000 ft above sea level. At Dimapur ran the Ledo-Calcutta railway, and the road represented the best route from Assam into Burma, while Dimapur was a vital railhead and supply base. Kohima was thus crucial to the Fourteenth Army, but initially Slim did not believe that the Japanese would make much of an effort against it, given the impossible nature of the terrain. The garrison, under the command of ex-Chindit Colonel Hugh Richards, initially consisted of some platoons of locally raised Assam Rifles and the raw Shere Regiment of the Royal Nepalese Army. Richards was also unable to obtain barbed wire from Dimapur because, he was reminded, there was a peacetime regulation forbidding its use in the Naga Hills.

The British 2nd Division, which had for many months past been carrying out training in the Belgaum area and had not seen active service as a whole, although four of its battalions had fought in the Arakan, was rushed by road, rail and air to defend Dimapur. Here they joined the 4th Royal West Kents, who had recently flown in from the Arakan as part of 161 Brigade of 5th Indian Division. The RAF reported that the Japanese were beginning to surround Kohima, and hence the Royal West Kents under Lieutenant-Colonel 'Danny' Lavery were rushed to it as the spearhead of 161 Brigade. The 1st Assam Regiment was also pulled in.

Lavery's men arrived on 5 April and on that same day the Japanese cut Kohima off from the outside world. Holding the ridge which dominated the road was a garrison of 2,500 men, including 1,000 non-effectives. They were faced by 6,000 Japanese. Confined to a triangular area measuring no more than 700 yards × 900 yards × 1100 yards, the defenders had plenty of food and ammunition, but water was in short supply. The only supporting artillery was a battery of mountain guns. For the next two weeks a bitter battle was to be waged for points such as 55 IGH (Indian General Hospital) Spur, GPT (General Purpose Transport) Ridge, DIS (Detail Issue Store), and FSD (Field Supply Depot), names which reflected Kohima's sleepy backwater existence before the Japanese struck.

Some of the most desperate fighting took place on the District Commissioner's tennis court, with the Japanese at one end and the British at the other. Nowhere within the perimeter was out of range of Japanese fire, and for the wounded, in particular, it was an appalling experience, with many being hit again as they lay. Indeed, after a while, in the words of one Royal West Kent officer: 'Many of the wounded, I feel sure, died in the last few days because they had given up hope.'

The situation became increasingly grim, with the remainder of 161 Brigade themselves isolated in a box at Jotsoma, some two miles west of Kohima, and the road to Dimapur also cut some three miles from Jotsoma. Finally, on the night of 17/18 April, the Japanese managed to cut the garrison in two by seizing a point midway along the ridge. Curiously, though, they did not immediately follow this up with a *coup de grâce*.

Yet those still on their feet never gave up. Their actions were epitomized by those of newly promoted Lance Corporal John Harman of the Royal West Kents. The Japanese captured the garrison bakery at one point in the FSD and hid in the ovens. A company counter-attack by the Royal West Kents met withering fire, but, covered by his section, Harman went forward alone and dealt with two machine guns which had been engaging from the flanks. A further attack still failed to regain the bakery and so, undeterred, he entered it himself with a box of grenades and methodically went round all the ovens, lifting their lids and dropping grenades inside them. Sadly, he was later killed, but his deeds were recognized by a posthumous Victoria Cross.

On 18 April, when all seemed lost, the 1st/1st Punjabis managed to fight their way through from Jotsoma, to be followed by the remainder of 161 Brigade. An eyewitness:

'The liberators saw little groups of grimy and bearded riflemen standing at the mouths of their bunkers and staring with bloodshot sleep-starved eyes as the relieving troops came in. They had not had a wash for a week. With the boom of the guns and the screech of shells always in their ears, they had fought and lived in their trenches for almost a fortnight. For rest they had thrown themselves on the ground with their boots on, ready to fight at a moment's notice.

It had been an epic, as fine as anything in the annals of the British Army. As Slim wrote: 'Sieges have been longer but few have been more intense, and in none have the defenders deserved greater honour than the garrison of Kohima.'

But the fighting here was by no means over and would continue for another two months. First 2nd Division still had to clear the road from

Dimapur and then advance itself into Kohima. Next it was a question of clearing the Japanese from their now well-prepared positions on the hills and ridges around the village as well as continuing to repulse attacks on Kohima itself. During this the Division had over 2,000 casualties, a measure of just how tough the fighting continued to be. At one point, after an attack, the 2nd Royal Norfolks, whose Captain John Randle won a posthumous Victoria Cross, found among the Japanese dead capbadges belonging to the men of three of their sister battalions who had been killed or captured on Singapore over two years before.

Around Imphal, as well, there was a hard tussle, and it was not until the Imphal-Kohima road was reopened on 22 June that the battle was finally won and the Japanese withdrew. The total British casualties for the three months were 17,500, while the Japanese lost 60,000. But, in the words of one Japanese officer: 'I have fought three times in my life: in the battle of the Shanghai line, in north China, and in Burma. Each time there was very heavy fighting . . . but the heaviest was at Kohima. It will go down as one of the greatest battles in history.'

While the battles for Imphal and Kohima were taking place, the Fourteenth Army was carrying out more offensive operations behind the Japanese lines. The first Chindit expedition had caught Churchill's imagination to such an extent that he took Wingate with him to the Anglo-US strategic conference held at Quebec in August, 1943. As a result Wingate was given authority to conduct another, but larger scale, long-range penetration in Burma. Accordingly he formed 3rd Indian Division (Special Force). This was made up of no less than six brigades and consisted of three Gurkha, fourteen British, and three West African battalions. The battalions which took part in the first expedition were not involved, although The King's Regiment was represented by its 1st Battalion.

The eventual plot hatched for the Chindits was that they would operate in conjunction with Stilwell's Chinese in order to make northern Burma safe for the Ledo-Kunming road, vital for keeping the Chinese supplied with munitions, to be completed. Stilwell would cut off the Japanese 56th Division in Yunnan, while the Chindits did the same to 18th Division in northern Burma.

The first Chindit brigade, Bernard Fergusson's 16th, set off on foot from Ledo on 5 February and began to make its way south towards Indaw. A month later two further brigades, 77 and 111, were flown into Burma, landing by Dakota and glider on two landing grounds previously selected by air reconnaissance. The former brigade suffered, however, during the fly-in because two gliders were attached to each Dakota, which proved too heavy a strain on the aircraft engines, and only thirty-five of the sixty-one gliders which set out actually reached the landing ground at Broadway.

Even so, by 13 March Wingate had 13,000 troops deployed. Ten days later 14 Brigade also began to fly in, the other three brigades having each set up base strongholds.

Matters now began to go wrong. Firstly Wingate went against his previous policy of not attacking fortified positions by ordering 16 Brigade, footsore after fifty days' march, to attack Indaw. The Brigade set out from the stronghold Aberdeen to traverse the 25 miles to its objective. Much of the terrain, however, turned out to be waterless, which served to add to the hardships that Fergusson's men had already suffered. Indaw proved too tough a nut to crack, even though the 2nd Leicesters, under their wounded colonel, clung on tenaciously to their positions close to the airfield for three days in the face of heavy Japanese counter-attacks.

In the midst of all this Wingate himself was killed in a plane crash, and then White City and Broadway, the other strongholds, were attacked. This was the second time that the former had been attacked, the first being while it was still being established. Then a fierce fight developed on Pagoda Hill, which involved the 2nd South Staffords and 3rd/5th Gurkhas having to retake it. An officer of the latter:

'So, standing up, I shouted out "Charge!" in the approved Victorian manner, and ran down the hill with Bobbie and the two orderlies. Half the South Staffords joined in. Then, looking back, I found a lot had not. So I told them to bloody well "Charge, what the hell do you think you are doing?" So they charged. Machine-gunners, mortar teams, all officers – everybody who was on that hill.'

The Japanese also charged, and the result was a hand-to-hand mêlée with 'everyone shouting, bayonetting, kicking at everyone else, rather like an officers' guest night'. At one point a young subaltern of the South Staffords had his arm chopped off by a Japanese officer's sword. Undeterred, the officer shot him with his revolver, picked up the sword and used it until collapsing. Then he whispered to his brigade commander, Mike Calvert, who knelt beside him, 'Have we won, sir? Was it all right? Did we do our stuff? Don't worry about me,' and expired.

Defending strongholds, however, was doing little to further the aim of the operation, and Calvert and Masters, commanding 77 and 111 Brigades, were ordered to leave them and advance northwards towards Mogaung. Fergusson's exhausted men were flown out, but Joe Lentaigne, Wingate's successor, did now have the West Africans of 3 Brigade, and they and Brodie's 14 Brigade also marched north. Being out of their strongholds meant, however, that air resupply and the evacuation of sick and wounded were more difficult, and Masters was ordered to occupy another stronghold,

Blackpool, some 20 miles south-west of Mogaung. This came under Japanese attack almost immediately. John Masters:

'Work was going on slowly as the men's exhaustion grew on them. The guns had come in with the gunners and the dozers were moving them into position . . . Everyone had been up most of the night, and now we had to carry the stores up from the strip, put down more wire, dig more positions deeper, repair others, bury the dead, patrol the jungle, protect the airstrip.'

The pressure continued throughout the second half of May as almost a complete Japanese division was deployed against Masters' 2nd Cameronians, 1st King's Own Royal Regiment and Gurkha Rifles. After a time it became impossible to keep him resupplied, and eventually he was forced to break out, thus enabling the Japanese to reinforce Mogaung. It was, however, this town which Stilwell, advancing from the north, expected the Chindits to capture.

One of the worst aspects of the break-out was dealing with the badly wounded, who could not march. Richard Rhodes-James: 'One man had been wounded in the leg and was dragging himself across the paddy towards the path along which we were moving. On looking back at this moment afterwards, I reflected how gallant it would have been to have rescued him. At the time, I was conscious only of the men pressing me from behind and the need to keep going.' Some even had to be shot by their comrades to prevent them falling into Japanese hands alive and thus laying them open to unimaginable tortures.

Masters and his men managed to link up with the West Africans and received much-needed succour. By now the Chindits had been operating behind the lines for well over the ninety days maximum which Wingate had considered was possible for them. Yet 'Vinegar Joe' Stilwell was insistent that they play their part by capturing Mogaung, and 77 and 111 Brigades, which had seen the heaviest fighting, were tasked to do this.

Calvert's men led the way, but Mogaung was now a fortress, and after his initial attempts, the 1st Kings were down to a mere seventy men, and combined with their fellow Lancashiremen, 1st Lancashire Fusiliers. But then some of Stilwell's Chinese arrived. More important, they had artillery, which Calvert lacked. On 24 June Calvert attacked again, his total infantry strength down to a mere 520 men. Yet his Gurkhas, South Staffords and men of Lancashire, supported by Chinese guns, still managed somehow to find hidden reserves sufficient to overrun the town, but at a dreadful cost in casualties. 'The Chinese were full of admiration, but thought we were quite mad, for with oriental patience they would have taken a week to do the same attack and probably suffered five per cent of our casualties,' wrote Lieutenant Durant of the South Staffords.

Yet Stilwell, under whose command the Chindits had now been placed, had not finished with them, and was demanding yet more, accusing them, when they protested at their diminished strength and exhaustion, of being malingerers. Doctors were sent to examine them. Such were the ravages of malaria, typhus, foot-rot, septic sores and prickly heat that in Masters' brigade a mere 118 were found fit for further action, or 119, since he insisted on including himself in the total. Even then, when Masters reported to Stilwell for orders, having evacuated his sick and wounded, he was ordered to take his now company-sized force to go and guard a Chinese medium battery. Only when the British 36th Division, whom we last heard of during the Madagascar landings and which was now also under Stilwell's command, had passed through were the remnants of 111 Brigade eventually lifted out. Those of 3 West African, 14 and 77 Brigades were not released by Stilwell until the end of August. It was a sad and frustrating end to an expedition which had been launched with such high hopes over six months earlier. Indeed, the contribution that the Chindits made to victory in Burma has been a subject of hot debate among historians ever since. Yet, whatever the rights and wrongs of the issue, there is no doubt that the Chindits themselves, the bulk of whom were made up of English and Scottish infantry battalions, displayed fortitude and endurance of the highest order.

At the beginning of December, 1944, the Fourteenth Army crossed the Chindwin on a broad front and began its offensive to regain Burma. In mid-January Slim's men crossed the Irrawaddy. This in itself was an achievement, since the river was up to 1500 yards wide. Corporal Dillon of the 2nd Border Regiment was a member of one of the reconnaissance patrols that first crossed the river to check on Japanese activity on the far bank in 20th Indian Division's sector:

'There were three of us on this patrol, Sergeant Harewood, Private Bennion and myself. We drew four days' rations and blacked our faces in dirty grease from the cookhouses. At about seven o'clock in the evening, before the moon rose, we shoved off in a rubber boat. Halfway across, some parachute flares lit everything up but we lay flat in the boat for fifteen minutes and nobody saw us. We got to the other side and let the air out of the boat and pulled it up to the shore very cautiously, crawling on our stomachs.'

Just as they had hidden the boat by a tree in a pea field a party of Japanese arrived and began constructing bunkers just ten yards from where they lay, and they had to crawl for an hour and a half to get out of the way, but again their way was barred. They therefore got up and bluffed their way across a track, sauntering across in full view of the sentries. They spent what remained of the night in elephant grass by a lake, only to have Japanese come down to a point five yards from them to water their mules. The second night they spent in a paddy field and then hid up by a road junction to monitor Japanese activity and managed to locate a supply dump in a village. Next morning they checked the high ground overlooking the river and moved that night six miles south. At daybreak they found signs of a Japanese column on a track and then located a gun position. 'By now our lips were cracking with thirst, so that night we got back to the paddy fields and got right down on our stomachs and drank the dirty water. Then we opened two tins of cheese, but they were both bad. That left us only sardines.' The following night they went back to where they had hidden their boat:

'When we got there ... we found a party of Japs still there with a sentry on each of the tracks. We were crazy to go in, it was a suicide job, but I think we were fuddled after five nights without sleep. Sergeant Harewood led us across the track, straight to the tree where the boat was hidden. A miracle that was. It took us an hour to get the grass off it without making a rustle. Then Brennan and the Sergeant dragged it down to the other track, inch by inch, and I went forward and gave a low whistle when the sentry's head was turned. We got it down to the water's edge, and we'd just blown it up when five Japs came down to the river, about ten yards away on our left. We rolled into the water and stayed there with out Stens ready for ten minutes until the Japs had filled their water carriers and gone away. Then another lot came down five yards away on our right, and went away again. So we didn't wait to blow the boat right up.'

Their troubles were still not over. The boat proved difficult to manoeuvre and they made only some forty yards across the river in an hour and a half. Eventually they reached the home bank and were welcomed by some Indian machine gunners, who had been covering them from midstream onwards, thinking that they were enemy. They 'came round and patted us on the shoulder and said "Tikh hai, Sahib!" Tikh hai is Hindustani for OK. We were tikh hai all right after the officer in charge had given us a cigarette and some rum.'

Preceded by air attacks on gun positions, the crossings were made with little of the sophisticated amphibious equipment used to cross the Rhine, and were largely reliant on using chains of ranger-boats with outboard motors. The 1st Northamptons, also in 20th Indian Division, had a tough time when theirs broke down and they drifted downstream past their planned landing point to one where the Japanese fire was heaviest. The British 2nd Division's landings very nearly did not succeed at all. Many of the 7th Worcesters' boats filled with water or were sunk and none made it to the far bank, while the 1st Royal Welch Fusiliers got no more than two platoons across. The 1st Camerons, on the other hand, managed to get just over a company across under heavy sniper fire, and after climbing a steep cliff on the far bank. It was, however, only when the 2nd Dorsets were switched from their landing point to that of the Camerons, which meant a march of some three miles, that a bridgehead was properly secured.

Further south, the 7th Indian Division crossings, which were part of the Mandalay deception, also had their problems. Selected to spearhead them were the 2nd South Lancashires, new arrivals to the Division, but who had taken part in the Madagascar landings and hence were expected to know something about amphibious operations. The plan was for one company to cross in the hours of darkness and for the remainder to follow at first light. The lead company got across with no opposition and began to dig in. The rest of the South Lancashires had orders not to start their outboards until they were in midstream. Some of the boats were found to leak and the motors of others would not start. Confusion grew and the carefully laid-down crossing order became disorder. Eventually all the boats, unwillingly and willingly, began to move downstream, only to come under heavy fire. Two company commanders and the engineer officer in charge of the crossing were killed and the commanding officer's boat sunk. Eventually aircraft had to be called in to subdue the enemy fire in order to allow the South Lancashires to get back to the home bank.

Meanwhile, the lead company completed its digging in, still without experiencing any opposition. The brigade commander, worried about their isolation, brought forward one of his Indian battalions, who crossed, and the bridgehead was properly secured. The Japanese did counter-attack a number of the bridgeheads, but these were beaten back and, after three weeks' operations, the thrust east and south could continue.

Meiktila was quickly seized, although again the Japanese, aware of its importance, counter-attacked fiercely. Then came the epic battle for Mandalay, which involved three divisions, 2nd British and 19th and 20th Indian. General Pete Rees was commanding 19th Indian, which included three British battalions (2nd Welch Regiment, 2nd Worcesters, 2nd Royal Berkshires). He transmitted some live and very graphic descriptions of the fighting on BBC. The mopping up of Mandalay Hill after the Gurkhas had reached its summit:

'We're sending up a company of the Royal Berkshire Regiment to assist. They're now on the way. Meanwhile, right at the bottom of the mountain, at the north end there's still some stubborn fighting going on. Japanese in machine-gun nests and anti-tank defences there, and at the moment Gurkhas and Baluchis, supported by British tanks are dealing with that. The rest of the Royal Berkshires are just behind me, and their last company is slogging in from the north, very dusty, in very good spirits. We had to send them back yesterday evening in the dust to clear a village near where I had my headquarters last night, when they kindly cleared it for me . . . Now here it's a very sunny, hot morning, it's very dusty, the troops are very dusty and they're pouring with sweat. For some days now they've been going all out. They've had a neglibible amount of sleep, but their spirit and *élan* is tremendous. I've never, never seen such enthusiasm and it's that, undoubtedly, that is carrying them through, because they feel the prize is really worth the effort.'

The battle for the last Japanese stronghold in Mandalay, Fort Dufferin, was especially tough. Pete Rees, again, describing one of the attacks on this bastion:

'You can hear the noise of the shelling, mortaring, shooting. I'm fairly close to the walls myself, standing, looking half round a concrete wall. Our chaps are advancing steadily, bunching a little more than I'd like to see them. They're going very well. The tanks are advancing, firing very hard at the walls . . . I can see the breach, but there's a big moat, this side. I can now see some of our leading infantry. They've just doubled behind a concrete shelter . . .

'Tremendous lot of noise going on. A whole lot of smoke now, near the wall itself, which is a very good thing for our infantry. I'm not sure which of the firing is the enemy firing. I can see some of our infantry running round the tanks. Not always wise to stand near a tank. Now I can see more of our infantry going across now, they're running across near the tanks, they're in slouch hats.'

Fort Dufferin fell on 20 March, and a week later the Japanese in front of Meiktila drew off.

Now began the thrust south to Rangoon, which was liberated on 3 May. This left the Japanese Twenty-Eighth Army trapped in the Arakan and under constant pressure from Christison's XV Corps, while the remainder withdrew eastwards into the Shan Hills.

The final battle of this long and arduous campaign was that of the River

Sittang. The Japanese Twenty-Eighth Army stopped at nothing in its attempts to break out of its trap and at times the Sittang literally seemed to run red with blood. In the midst of this the monsoon was raging, adding to the discomfort. Steps were also being taken to mount a major amphibious landing on the Malayan coast, but the end of the war came before this had to be mounted in anger.

The assertion has been made that the British soldier of 1939–45 lacked the moral fibre of his father's generation. One of the main planks of this argument is that the infantry battalion of 1914–18 endured greater casualties, and yet remained in being as an effective fighting force. True, the soldier of the Second World War had much higher expectations of life, but this was inevitable, given twenty years of increased education and the growing sophistication of life brought about by advances in technology. This was reflected in a much greater emphasis on troops' welfare than in the Great War. Yet, when it comes down to it, the 1939–45 infantryman often endured just as severe conditions on the battlefield as Old Bert and Young Bill had in the mud of Ypres, but continued to do his job. The main difference was that his senior commanders, all of whom had endured the trenches and experienced their slaughter, were determined that their men should not undergo *needless* suffering. Therefore, they would do their utmost to relieve battalions after heavy fighting sooner rather than later. Besides, the spectre of the manpower crisis loomed earlier and more starkly than it had during 1914–18.

CHAPTER TEN

Postwar Turmoil
1945-1949

DURING the course of 1939–45 the number of infantry battalions in the British Army had grown from 140 to 680. Significantly, though, and a reflection of the growing complexity of land warfare, the proportion of infantry within the Army as a whole had fallen from one fifth to roughly one seventh. Demobilization, which operated under a much more efficient and fair system than in 1919, resulted in the number of battalions being reduced to 143 by March, 1947, with the Territorial Army being suspended until the Government had worked out a way forward for its postwar defence policy.

In the meantime the immediate problems of the postwar world kept the British Army at full strength. Beginning in the Far East, the British had to contribute to the occupation forces in Japan, and a division, largely Indian, was formed for this. It included one of the brigades of 2nd Infantry Division, made up of the 2nd Royal Welch Fusiliers, 2nd Dorsets and 1st Camerons. But by the end of 1948, because of commitments elsewhere and the granting of independence to India and Pakistan, the contingent had been withdrawn.

Another Indian Division, Douglas Gracey's 20th, was sent to Indo-China to disarm the Japanese and hand the country back to the French, but the only British unit involved was a gunner regiment. A similar task confronted other Indian formations in the Dutch East Indies. This, like Indo-China, proved to be a more difficult operation than was first envisaged. As the indigenous population of Indo-China had no wish to see the French back, and at one point the Indian troops found themselves operating with the Japanese against the Vietminh, so there was a wave of nationalism in Java and the other islands. The Dutch colonists, even though they had been freed from their camps, viewed their liberators with suspicion, fearing that they intended to take the Dutch East Indies for themselves. Perhaps they remembered the last time that the British had come to Batavia, during the Napoleonic Wars. The result was, as a report by the 1st Seaforths put it: 'The general attitude towards the Dutch is not one of great friendliness. This may be partly due to a lack of understanding of what the majority of Dutchmen out here have been through during the last three years, but is greatly influenced by the often rude and hostile attitude of the Dutch themselves.'

Even so, the colonists had to be protected against the increasing stridency of the nationalists, as had the Chinese population, who also came under threat. By the end of October the British found themselves with a full-scale insurgency on their hands, their opponents making much use of captured Japanese weapons, and indeed, joined by escaped Japanese POWs. At one stage the British had to employ ground attack aircraft and naval gunfire. By Spring, 1946, however, the major population centres were reasonably secure, and, as Dutch troops arrived from Europe, the British were thankfully able to begin to withdraw from a distasteful proxy operation which had cost them some 2,000 casualties.

In terms of British colonial possessions, there was no initial resistance to the British return to Malaya and Singapore; that would come later. Burma, however, proved more problematical. Not only had the infrastructure of the country been ruined by the war, and there were some 70,000 Japanese POWs to take care of, but there was a resurgence of the traditional Burmese pastime of dacoitry. Furthermore, the Burmese made it clear very early on that they wanted independence from Britain. Matters were not made any easier by the fact that there were two Burma armies, one trained by the British and built round the Burma Rifles, and the other formed by the Japanese. It was Aung San who was the power behind the latter, and who made himself the main focus for nationalism. Luckily, he proved to be more moderate than expected.

At the end of hostilities in Burma the large Allied forces had been rapidly run down, and by Autumn, 1946, just fifteen British and Indian infantry battalions remained, besides supporting arms and services. The British element consisted of the 4th Border Regiment, which was disbanded in early 1947 and replaced by the 2nd Durham Light Infantry from Singapore, the 2nd Welch, 2nd Worcesters and 2nd Royal Berkshires. Their duties varied from providing guards for key points and supervising the Japanese POWs, including their repatriation in 1947, to drives against the dacoits, who became increasingly infiltrated by the communists.

'The van in which the guard was travelling stopped in a very steep cutting . . . and withering fire from rifles, Bren and other automatic weapons was brought to bear on them. From the accounts of those civilians who escaped with their lives, the guard fought very gallantly for over an hour against an overwhelming superiority of arms and numbers before they were eventually killed. This deliberate attack on men of this unit awakened a desire in all ranks to track down this band of dacoits (which was reported to be over 300 strong) and accordingly plans were laid.'

The Welch were successful in tracking down some of the perpetrators, and the remainder were accounted for in the highly successful Operation FLUSH, which involved six battalions, including two Burmese and the 2nd Durham Light Infantry. Yet, in spite of fears that Burma might boil over, the peaceful transfer of power was achieved, notwithstanding the assassination of Aung San in July, 1947, and resultant threat of a civil war, and the last British battalion, the 2nd Royal Berkshires, left on 3 January, 1948, thus ending a British military connection with the country which had begun in 1826. It is significant that the 2nd Welch should have been one of the last battalions stationed in the country, since one of their predecessors, the 41st Foot, had been among the first.

After the unrest of 1942 the British authorities had managed to keep the lid on Indian nationalism for the remainder of the war. By early 1946, however, the country was beginning to seethe once more. Matters were not made any easier by growing discontent among the British forces. In 1944 a scheme called PYTHON had been introduced, whereby those who had served overseas for four years or more had the right of repatriation to Britain. This began to take real effect shortly after hostilities came to an end. At the same time it was made clear to those who had been conscripted during the second half of the war that their service would not be terminated before 1947 and that later arrivals could expect to serve well beyond this. The loss of many experienced men and discontent among the newer conscripts that their return home was likely to be considerably delayed resulted in a rapid lowering in morale. In two cases this became open mutiny. In May, 1946, over 250 men of the 13th Parachute Battalion, recently arrived in Johore from the Dutch East Indies 'downed tools' in protest at the primitive conditions of the camp that they found themselves in. They were subsequently court-martialled and sentenced to long terms of imprisonment, although these were later quashed on the grounds of irregularities in the court proceedings. Earlier, in January, there had been another strike by the RAF in Calcutta, although this was settled through negotiation. However, the infantry battalions in India and the Far East appear to have been successful in containing troubles of this nature, but the concern remained that the discontent might spread.

Unrest in India was further aggravated by the trial in Delhi of three officers of the Japanese-sponsored Indian National Army on charges of waging war against the King-Emperor and abetting murder. The Congress Party used this to fan the flames and there were mutinies by the Indian

15. A corporal hunting a German sniper who has just killed his companion, Anzio, January, 1944.

16. The breakout from Normandy. On the left is a Bren-gun carrier mounting a Vickers medium machine gun from a divisional machine-gun battalion. The infantry section is led by a Bren gunner. The third man back is armed with a Sten gun.

17. Street fighting in Germany, early 1945.

18. Chatting up German girls just after the policy of no fraternization with the Germans had been relaxed in July, 1945.

19. The scene of the Glosters' epic stand against three Chinese divisions in Korea, April, 1951.

20. Manning an American 0.30-inch Browning machine gun in Korea.

21. Part of the Hook position in Korea the morning after one of the numerous Chinese attacks against it.

22. 1st Somerset Light Infantry in the Malayan jungle, mid-1950s.

Navy and Air Force. Rioting broke out in the main cities and it seemed that the country was about to disintegrate into anarchy. The situation became calmer when a British Government mission was sent to India in March, 1946. Their message was that India should have her independence, and it was only a question of the timing. In the meantime elections had been held, but had revealed that the country was split down the middle between the Hindu Congress Party and the Muslim League. The latter became insistent on the creation of its own state of Pakistan and, in August, 1946, matters degenerated once more with a wave of internecine violence. This was especially bad in Calcutta, where in just three days 20,000 people lost their lives or were seriously injured. At one point the 2nd York and Lancasters and 2nd Green Horwards had to resort to using tanks of the 43rd Royal Tank Regiment, equipped with powerful searchlights, to help regain control of key points in the city. There was, however, nothing they could do to stem the violence in the back streets.

The arrival of Mountbatten as Viceroy in February, 1947, with the mission of accelerating independence caused the focus of violence to switch to the Punjab, through which the boundary between India and Pakistan was to be drawn. Here the Punjab Boundary Force was set up under General Pete Rees in order to try and keep the peace. It was a mere two divisions strong, with a mixture of Indian and British troops, and had an area of responsibility larger than the whole of Ireland, with some fifteen million people, equally split between Muslim and Hindu. Trying to keep the peace between the two was an impossible task, especially in the knowledge that indiscriminate use of force would merely inflame the situation still further. Once the boundary between the two new states had been announced in mid-July panic set in as literally millions took what possessions they could carry to try and avoid being caught on the wrong side of the border when independence came into effect, which it was to do at midnight on 14–15 August.

The massacres grew, and the British troops were withdrawn to barracks. The Punjab Boundary Force did not, however, last long after independence. Both sides accused it of favouring the other and, at the end of August, it was disbanded, the new nations taking responsibility for their own side of the border. This did nothing to stop the killing, which increased still more. As it did so the British troops began to leave for home, those remaining merely having the responsibility for protecting British citizens and their property. Last to depart were the 1st Somerset Light Infantry. On 28 February, 1948, they paraded in front of the traditional Gateway of India in Bombay. During this farewell parade, which was conducted in the presence of guards of honour drawn from four battalions of the new Indian Army, the Somersets were presented with a silver model of the Gateway, inscribed 'To commemorate the comradeship of the soldiers of the British and Indian Armies 1754–1947'. Then, to the strains of *Auld Lang Syne*, and still proudly bearing 'Jellalabad' in their cap badges in memory of the Illustrious Garrison they trooped their Colours through the Gateway and thence embarked on their troopship for home.

Yet the British Army's connections with the Indian Army were not totally severed. A number of affiliations were formed with Indian and Pakistani regiments and, more significantly, one element of the old Indian Army was transferred. Throughout much of the eighteen months preceding independence delicate tripartite negotiations had taken place among the British, Indian and Nepalese authorities over the possibility that Gurkha regiments might be incorporated in the British Army. This was especially urgent since the Indians had made it clear that they were only interested in maintaining a portion. The Army Council was at first concerned that this could only be done at the expense of British infantry battalions, and there was already a plan, of which more later, drastically to reduce the number of Regular infantry battalions. However, the need to maintain a force base in the Far East proved overriding as an argument, and it was eventually agreed that two battalions each of the 2nd, 6th, 7th and 10th Gurkha Rifles were to transfer. In early 1948 they moved from India and Burma, where three of the battalions had been stationed, to Malaya. Here they formed 17th Gurkha Division, which also included three British battalions – 1st Buffs, 1st King's Own Yorkshire Light Infantry and 1st Seaforths – who were already in Malaya. Very soon the Division was to find itself bearing the brunt of the first phase of the Communist insurgency.

Moving to the Middle East, the end of the war found British troops stationed in Libya, Egypt and Palestine. In the last-named there was trouble brewing, and, indeed, very close to the surface. This time it was not the Arabs who were discontented, but the Jews. Even though the British had successfully protected them during the pre-war Arab revolt, there were some Jews who remained convinced during 1939–45 that Britain, rather than Germany, was the true enemy since it was denying them an independent state. These hardliners formed terrorist groups, of whom the most notorious was the Stern Gang, and carried out attacks and sabotage against the paraphernalia of British administration. Immediately after the war the British, in an attempt to be even-handed, continued to restrict the flow of Jewish immigrants, even though these were now mainly surviving victims of the Nazi Holocaust. They also announced that they were giving up the plan of separate Arab and Jewish states in Palestine since this would be impossible to achieve. Jewish anger turned to violence and by Spring, 1946, the Army was forced to commit two complete divisions, 1st Infantry and 6th Airborne, to Palestine.

Matters went from bad to worse. What had begun as rioting and sabotage attacks degenerated into assaults on British troops. At the end of April, 1946, seven members of the 5th Parachute Battalion were murdered in their tent on the waterfront of Tel Aviv, and two months later five officers were kidnapped in order to force the authorities to rescind the death sentence on two terrorists. The climax came with the blowing up of the King David Hotel in Jerusalem in June, with the loss of over ninety lives, British military and civilian. The immediate result of this was a massive cordon and search operation in Tel Aviv, involving 17,000 troops, which netted many illegal weapons and led to the arrest of several terrorist suspects. It became increasingly clear to the British Government, however, that the violence could only be contained if many times the existing force was deployed. Trying desperately as it was to cut down on defence commitments, and with ever less hope of finding a political solution to the problem, the Government eventually decided, in April, 1947, to hand Palestine over to the United Nations.

In November of that year the UN adopted a proposal of two separate states, with Jerusalem being given international status. The British mandate was to continue until 15 May, 1948, and all British troops would be out of the country six weeks later. This served to intensify the enmity between Arab and Jew, as both jockeyed for territory, and by Spring, 1948, the British garrison was being shot at by both sides. The troops were therefore withdrawn into enclaves based on Jaffa, Jerusalem and Haifa. During this operation some small detachments had literally to fight their way out. This happened to two platoons of the Irish Guards and a troop of the 17th/21st Lancers at Safad in north-east Galilee, which had been entirely surrounded by Arab forces. Within the enclaves the principle of minimum force began to be ignored as the British troops became more desperate. Thus, when Jews seized the house of a prominent Arab in Jerusalem and refused to evacuate it, the 1st Highland Light Infantry, supported by tanks and armoured cars, were sent in to remove them by force and inflicted a significant number of casualties.

For the final withdrawal itself, additional reinforcements, including two Royal Marine Commandos, were sent in. The Jerusalem garrison, which included 2 Guards Brigade, conducted a very skilful operation, including having to force its way through the Latrun Pass on its way to the port of embarkation, Haifa. As the troops withdrew, so open war between the infant state of Israel and her Arab neighbours erupted. The 30 June deadline was, however, met and none were sorry to leave after what had been a thankless task.

On the other side of the Mediterranean British troops spent the whole of 1945 helping the Greek rehabilitation after the civil war that had erupted at the end of the previous year. In early 1946 hostilities between the communistis and democrats broke out once more, but this time the British Government, determined not to be committed to sending more troops, quickly passed the problem over to the Americans, and withdrew its forces without them becoming embroiled.

The Balkans also gave rise to another potential conflict. Marshal Tito claimed the port of Trieste, which roused the Italians to fury. British troops had arrived there in May, 1945, at the same time as the Partisans and for a time it looked as though hostilities might break out. Eventually the United Nations designated it a Free Territory, but British troops remained there until 1956 as part of the monitoring force.

In Austria the Eighth Army faced a different problem in the immediate aftermath of victory. The prime task was rounding up the vast numbers of surrendered enemy troops. All had to be screened, as they did in Germany, for members of the SS, who were placed in the 'arrest on sight' category, and could usually be identified by their distinctive blood group tattoo near the left armpit. While most surrendered soldiers were apathetic to their fate, there were some Nazi diehards. An officer of the 1st Royal Irish Fusiliers, whose battalion captured a Hitler Youth battalion:

'These were fifteen- or sixteen-year-old boys who had sworn to take an Allied soldier with them before they went. They did surrender, but we had great difficulty in guarding them, because they were bent on mischief. It took one platoon all its time to look after a smallish battalion, and the men were on "two hours on, four hours off" on a twenty-four-hour basis. We couldn't keep this up because a lot of the men were young National Servicemen and they were falling asleep over their weapons. We didn't dare relax with these youngsters, because we searched them and found that quite a number were still armed. We collected quite a tidy pile of weapons. It was really more tiring than fighting a war.'

Other groups of prisoners were even more problematical. The Irish Brigade was warned on 14 May that two large groups of Croats, totalling some 200,000 men, and escorting half a million civilians, were making for the British lines, intending to seek sanctuary from Tito's men. The latter, too, had a sizeable force in the area and it was clear that a very nasty battle was likely in the very near future. Unfortunately, communications with HQ 78th Division were tenuous and it was not possible to obtain a quick answer, which would have had to come from Field Marshal Alexander's headquarters, on what should be done. Consequently Brigadier Pat Scott had to act on his own. It was clear to him that if the Croats were allowed to remain inside the British area of responsibility it would be impossible to feed them. Furthermore, they had, after all, been fighting on the German side. Tito's men, on the other hand, warned him that they were about to attack the Croats. Scott, using a mixture of bluff and diplomacy, managed to persuade the Croats to surrender to the Titoists, who promised to treat them with humanity. Only subsequently was it discovered that the majority of the Croats were done to death, but, given the situation at the time, there was little else that Scott could have done.

Even more difficult was the Cossack problem. Some 40,000, including women and children, surrendered to the British. Again, they had fought on the German side and it had been agreed with Stalin at the 1944 Yalta Conference that they should be returned to the Soviet Union, that is provided they were Soviet citizens. The problem also landed in the lap of 78th Division, whose members were initially ignorant of what had been agreed at Yalta. The matter was complicated by the fact that a Bulgarian division, fighting with the Red Army, was in the neighbourhood of one of the Cossack groups and demanded that the Cossacks surrender to them. This they refused to do, having a very good idea of what might happen to them. They were therefore allowed to surrender to the British, with those on the spot guaranteeing their safety.

When the troops were told of what was in store for the Cossacks they were aghast, from the Divisional Commander downwards, especially since they seemed to be such attractive people and British officers had given their word on the Cossacks' safety. It was made clear, however, that they were to be handed over to the Russians and that the order had come from Churchill himself. The Cossacks got wind of what was afoot and in many cases resisted being moved out of their camps. Major McGrath of the 5th Buffs, whose company was ordered to move one group to a railway station in order to board trains for the Russian sector, related that it quickly became apparent that the Cossacks had no intention of moving.

'I ordered four three-tonners to back up to them and with about twenty men tried to get them into the vehicles. The wailing increased and a number indicated that they wanted to be shot by us rather than be sent to the USSR. With great difficulty a few were forcibly put on one of the trucks, but it was impossible to prevent them from jumping off, which they did. It appeared that certain men were the ringleaders of this sit-down strike, and as an example, I ordered four men to put one of these ringleaders on a vehicle. However, he created such a disturbance that I was forced to hit him over the head with an entrenching tool handle.'

In other cases the troops had to fire over the heads of the Cossacks, and in one there was a flamethrower demonstration to persuade the Cossacks to come out of their tents. Many soldiers were sickened over what they had to do, especially when they became aware that the Cossacks were right; the Soviet intention was to liquidate them. Lieutenant-Colonel 'Bala' Bredin, commanding the 2nd London Irish Rifles:

'I think there was clear evidence that shootings took place. Anyway, my men were quite sure of it and we heard enough about it to be sure that it was not the sort of thing that we ought to have a hand in. We also got reports that there were suicides on the trains, people jumping off or cutting their throats with bits of broken window pane. A fairly serious situation arose with the soldiers very nearly getting to the point of saying, "Sorry, sir, we won't obey your orders. We will not take these men to where they are just being mown down without any sort of trial".'

Bredin warned his superiors about the feelings of his battalion and it was quickly relieved by another. The task of repatriation was completed, but it left a sour taste in the mouth, and the matter is the subject of hot debate to this day.

British troops were to remain in Austria, as part of the Allied forces of occupation, until July, 1955, when the peace treaty formally ending the Second World War as far as Austria was concerned was signed. The last battalion there, 1st Middlesex, was, however, on its own during the final two years of the occupation. In contrast, British troops have remained in Germany ever since 1945, although from 1955, after West Germany had been admitted to NATO, as members of the Alliance rather than forces of occupation.* Initially, after the defeated German forces had been disarmed and disbanded, the main task was, as in Austria, to ensure that the German people did not starve and to set up a civilian infrastructure through which they could help themselves as much as possible. There were, too, enormous numbers of Displaced Persons from all over Europe, many victims of the concentration camps, to look after, and the battlefields to be cleared, especially of unexploded munitions. Vital points also had to be guarded against the threat of the Werewolves, a shadowy group of fanatical Nazis who swore to take their revenge on the victorious Allies, although in the event there were hardly any attacks mounted by them.

All this meant that life was very busy, although the setting up of leave

* Berlin was an exception. The British Berlin Brigade, together with its American, French and Soviet counterparts, remained as an occupation force until the Berlin Wall came down at the end of 1989. It was finally disbanded in 1994.

centres in the Alps and elsewhere did enable the troops to enjoy some of the fruits of victory. Initially, Montgomery established a strict 'no fraternization' rule, but this was dropped in July, 1945. Apart from the realization that it would make the Germans less likely to co-operate with the British authorities, it was also hard for the British soldier, who has a strong sense of humanity, to rigidly observe it, especially where pretty girls and children were concerned.

The role of the British Army of the Rhine (BAOR), as the British Liberation Army became known in November, 1945, was essentially that of a constabulary. As the months rolled on, however, it became clear that not only was the Soviet Union proving more and more difficult over co-operation with the Allies, but her attitude was becoming increasingly aggressive. Concern grew that Stalin not only had designs on Eastern Europe which ran counter to what had been agreed among the Allies, but on West Germany, and Western Europe as a whole. One major indicator of this was that while the Western Allies rapidly reduced their armed forces, the Soviet Union did not. Consequently, in early 1947 it was decided that two divisions in BAOR should be given an operational role. The 7th Armoured Division, which had been in Germany since the end of the war, was selected as one, and the other was 2nd Infantry Division. The latter had absorbed 36th Division in the Far East shortly after the end of hostilities and had then been sent to Malaya. At the beginning of 1946 it was, however, disbanded, but immediately resuscitated in BAOR, with the headquarters and unit cadres being sent from Malaya to help reform it.

Throughout the immediate postwar period the British Government had been trying to formulate a defence policy, but until the turmoil in the immediate aftermath of the war had receded it was impossible to do so. Originally it wanted to return to all-volunteer forces, but with the global commitments now facing Britain's armed forces this proved impossible and it was accepted that conscription would have to be maintained. Initially the Government wanted conscripts to serve for one year only, but when Slim took over as CIGS in November, 1947, he managed to get this increased to 18 months on the grounds that it would be impossible to send conscripts to the Far East, since, by the time they had completed their basic training and arrived it would almost be time for them to return home. It was not, however, until 1 January, 1949, that National Service, as it was called, came into effect. Until then the Armed Forces had remained reliant on the wartime call-up system. Every 18 year-old, provided he was physically fit, was liable, although those undergoing higher education could gain deferment until they had achieved their first degrees. For the next decade it was to be on the shoulders of the National Servicemen that the British Army would wage a number of campaigns.

The Treasury was insistent, however, that defence costs must be restrained through instituting a strict manpower ceiling. For the Infantry this meant a dramatic decrease in the number of Regular battalions. The 143 battalions existing in March, 1947, were reduced during the next year by a further fifty-two. This meant that, apart from the Foot Guards, all regiments were reduced to a single Regular battalion. Many battalions were placed in what was called 'suspended animation', but eventually those who escaped this were merely amalgamated and where 1st battalions had been suspended the 2nd battalion became the 1st. As for the Foot Guards, the Scots Guards lost their third battalion, and the Irish and Welsh Guards were reduced to a single battalion each.

This was a reflection that the Cardwell System could no longer cope, especially with the need to deploy reserves overseas at short notice. Indeed, this had aleready begun to be accepted. In October, 1946, the decision was taken for the purposes of recruit training that the infantry regiments were to be grouped as lettered brigade training groups as follows:

A –   Foot Guards
B –   Lowland
C –   Home Counties
D –   Lancastrian
E –   York & Northumberland
F –   Midland
G –   East Anglian
H –   Wessex
J –   Light Infantry
K –   Mercian
L –   Welsh
M –   North Irish
N –   Highland
O –   Green Jackets

Initially the training was run by the wartime Primary and Infantry Training Centres. These were finally abolished in 1948, and Regular battalions took it in turns to fulfill the training function for each brigade, thus avoiding further disbandments which might have meant the total loss of some regiments from the order of battle. In the same year, 1948, the Gurkha Regiment, as it was called on transfer to the British Army, became Training Brigade Group P. At the same time, the brigade concept was taken a stage further by making the brigade training groups into Regional Brigade Groups. The implication of this was that an officer or man could be posted to any regiment within the regional group. The one exception was the Foot Guards, who continued strictly to observe capbadge loyalties.

While regional brigade groups made some administrative sense in that they ensured that battalions on operational tours overseas could be kept topped up with men, there were many who saw it as the beginning of the end of the regimental system which had served the Infantry so well. However, during 1951–2, because of operational demands, the policy of deploying Regular battalions in the training role was done away with and there was reversion to the old regimental depot concept, something which gave hope to those who feared that the ultimate aim was to develop a corps of infantry. All the Cardwell regimental titles were retained, the only change being that in 1946 three regiments were awarded the 'Royal' prefix for their distinguished services during the war. These were the Lincolns, Leicesters and Hampshires, but only the Lincolns exchanged their white facings for royal blue, perhaps realizing that their laundry bills would be cheaper.

In 1947 the Territorial Army was reconstituted, but reduced in size compared to pre-1939. Now it was to consist of two armoured divisions (49th, 56th) and six infantry divisions (42nd, 43rd, 44th, 50th, 51st/52nd, 53rd). This meant a significant reduction in the number of TA infantry battalions, although each Regular regiment still retained at least one. In 1950, in order to ensure that former National Servicemen could act as an effective reserve of manpower, a system was instituted whereby they were placed on the books of the TA for four years and were expected to attend annual camp with their local unit in order to keep their military skills up to date.

When Montgomery became CIGS in early Summer, 1946, one of his main aims was to improve the soldier's lot. In August, 1946, he had argued that the soldier was 'part of the fabric of the nation' and was thus entitled to a 'good life'. This meant improving his comfort through getting rid of the barrack room in favour of bedrooms and sitting rooms. His food should be equal to that in civilian life, and irksome minor restrictions such as 'Lights Out' should be lifted. Given Britain's defence commitments and that money was in short supply, it proved impossible to make much progress in this direction. There was little money to spend on barracks and many troops had to endure increasingly delapidated wartime hutted camps. For the National Serviceman this often proved a rude shock after the comforts of home life. Indeed, one of their main memories was the barrack room in winter and the efforts to try and keep the coke-burning stove, the only source of warmth, alight. In spite of food rationing, much effort was made to ensure that the troops were well fed, and the fourth meal, supper, was now established as a proper meal, rather than merely consisting of the day's left-overs. Nonetheless conditions were hard, reflecting the country's austerity economy. Apart from the attempt to improve the supper meal, perhaps the only tangible results of Montgomery's efforts to enhance the quality of life was that in 1949 pyjamas became part of the initial clothing issue and that all ranks, not just officers, now wore a collar and tie with Battle Dress.

CHAPTER ELEVEN

The National Service Era
1948–1960

SOME of the first batch of National Servicemen were to find themselves on active service as soon as they had finished their basic training. On 16 June, 1948, three Chinese went by bicycle to the Elphil Estate, 20 miles east of Sungei Siput in Malaya, and murdered the estate manager. On the same day two further British planters met their deaths in similar fashion, and on the 17th the Government declared a state of emergency in Perak and Johore, to be extended the following day to the whole of Malaya. This state of emergency was to last for twelve long years. The brunt was initially borne by 17th Gurkha Division, the Malayan Police and the locally raised Malay Regiment, but in September, 1948, reinforcements arrived from Britain in the form of 2 Guards Brigade. This was significant in itself, as it was the first time ever that the Foot Guards had been deployed to the Far East.

The enemy was the Chinese Communist Terrorist, or 'CTs' as they were called. Most had fought against the Japanese in the Malayan jungle and had retained many of their often British supplied weapons. Led by Chin Peng, the Malayan Races Liberation Army (MRLA), as it became known in 1949, consisted of upwards of 10,000 men, well organized and equipped. The MRLA was also very disciplined, with a large fringe body, the People's Movement, which kept the terrorists supplied in their jungle bases and gathered intelligence. They were therefore a formidable enemy. As for their strategy, this was to be in three phases. First, they intended to establish liberated areas around their jungle camps through the murder of all Europeans in the vicinity. They would then join these up to establish a 'Liberated Country'. Finally, they would turn their attention to the urban areas, bringing about a breakdown of law and order, and disruption of the economy so as to create the political climate necessary for a Communist government to take power.

The CinC Far East, General Sir Neil Ritchie, was a firm believer that attack was the best form of defence and laid down from the outset that the

CTs must be harried wherever possible. One early innovation to this end was Ferret Force. Picked men were drawn from all battalions in Malaya and trained and equipped to operate for extended periods in the jungle with the aim of tracking down the terrorist camps and destroying them. In the main, though, operations were on a large scale. The two usual types were cordoning off a jungle area and sending in a force to flush out any terrorists within it, and to use a line of 'beaters' to drive the CTs into pre-positioned ambushes. It became increasingly apparent that these types of operation were too cumbersome and enabled most of the terrorists to slip away before they were trapped. Also, a number of the laboriously learnt skills of living and fighting in the jungle had been forgotten since 1945 and had to be reacquired. As it was, it was felt that men lost their alertness after ten days in the jungle and the general consensus of opinion was that smaller operations, acting on good intelligence, would be more effective. Worthwhile intelligence, however, could only be gained when there was close co-operation between the army and police, and in the early days mutual suspicion often prevented this.

The tide slowly began to turn after the appointment of General Sir Harold Briggs as Director of Operations in April, 1950. To ensure close co-operation among government, police and military, he organized operational and intelligence committees, on which all three were represented, at every level of command. He also aimed to remove the water in which the terrorist fish swam, to use Mao Tse-tung's well known dictum. Villages near the jungle's edge were relocated in order to prevent terrorist infiltration and intimidation, and much emphasis was placed on disrupting food supplies to the CTs. These measures forced the terrorists to operate in smaller groups, but did not prevent one of their most successful coups, the ambushing of the High Commissioner, Sir Henry Gurney, in October, 1951. He was replaced by Sir Gerald Templer, who took over as Director of Operations as well. He continued Briggs' strategy, but also laid great store on winning and maintaining the indigenous population's support for the security forces, what was called 'hearts and minds'.

While all arms served in Malaya, and, indeed, much use was made of the RAF, both to attack terrorist camps and to keep troops in the jungle supplied, and, to an extent, the Royal Navy, the main burden fell on the infantry. The Gurkha battalions, alternating postings between Malaya and Hong Kong, bore the brunt, but there were always a number of British battalions involved, as well as Royal Marine Commandos, Australians, New Zealanders, East Africans and Fijians, not to mention the Malay Regiment, Police and trackers from Borneo and Sarawak. Not surprisingly, since all their battalions served at least two tours in Malaya, the Gurkhas achieved the highest successes in terms of kills and captures, but some

British battalions also performed extremely well. The record was held by the 1st Suffolks, who served in Malaya from mid-1950 until early 1953, and accounted for 195 terrorists, themselves losing twelve killed and twenty-four wounded. One of their most notable successes was achieved by a National Service subaltern after six days' searching of the Kuala Longat swamp for a notorious CT leader, Liew Kon Kim. His patrol suddenly came across three armed terrorists, who turned and ran. Second Lieutenant Hands immediately shot one with his submachine gun, and set off after the other two. He overtook one of these, who turned out to be a woman, but armed, and shot her too. Even though now exhausted, he continued to struggle through the swamp and eventually ran the third to earth, despatching him as well. Dragging the body out of the swamp, he recognized it as that of Liew Kon Kim. For this sterling performance Hands was awarded a much deserved Military Cross.

Another notable achievement was that of the 1st Royal Hampshires in December 1955. They received Police Special Branch information, which they had obtained from a surrendered terrorist, that a group of terrorist leaders were attending a political indoctrination course in a camp in Selangor. A company immediately set out and attacked the camp, killing eleven out of the twelve occupants. The one survivor was captured a few days later.

It was operations like these which helped to break the back of the insurgency. Food denial and the highly successful 'hearts and minds' campaign also made significant contributions. Yet, as late as summer 1956 there were no less than twenty-four battalions in Malaya, including eight British. Nevertheless, a year later, on 31 August, 1957, Malaya received its independence from Britain, and an amnesty declared by the new government increased the growing number of MRLA desertions. During this last phase the Malayan Armed Forces were greatly enlarged and began to take over the main responsiblity for operations. By late 1959 the Loyals were the only British battalion left on an emergency tour and they departed at the end of the year, although 17th Gurkha Division remained in-country after the end of the Emergency was officially declared at the end of July, 1960.

On 25 June, 1950, an international crisis erupted elsewhere in the Far East when the North Koreans crossed the 38th Parallel in a lightning strike on their non-Communist neighbour, South Korea. The latter appealed to the United Nations and a resolution was passed to give active support to South Korea in repelling the invasion and restoring peace and security. Eventually seventeen nations were to send troops, while forty-nine sent supplies.

The first to arrive were US troops from Japan, and they were followed shortly afterwards by two British battalions from Hong Kong, the 1st Middlesex and 1st Argylls. They were grouped with the 3rd Royal Australian Regiment in 27 Brigade and, as such, fought their first actions in defence of the Pusan perimeter, the south-east corner of the country, which was all that remained in UN hands after the initial North Korean drive. They then took part in the break-out which immediately followed the successful amphibious landings at Inchon near Seoul.

It was during this that Major Kenny Muir, Second-in-Command of the Argylls, won a posthumous Victoria Cross for his leadership and gallantry during a two-company attack on a North Korean hill position, continuing to fight on even though he was mortally wounded. During the subsequent advance through North Korea, 27 Brigade, which was lacking supporting arms, was given a slice of American armour and guns. The Argylls at one point advanced so fast that, in the words of their Commanding Officer, Lieutenant-Colonel Leslie Neilson, who was looking for somewhere to spend the night:

'I became aware that we were no longer alone. Advancing towards us in single file on each side of the road were a large number of soldiers, stretching back along the road as far as I could see. They were clearly not Americans and there were no South Koreans within miles; they were in fact North Koreans, driven before the advancing US 24th Infantry. Small things stick in one's mind and I remember thinking how long their bayonets were; they had them stuck on their rifles for some reason. As soon as they saw us the first few files brought their rifles down and had a bang at us, but I had the presence of mind – or something – to shout to our chaps not to shoot. To try and bluff it out was the only way. This proved to be rather a puzzlement to the other side who, as it turned out – they were all put in the bag by the Australians – came to the conclusion that we were the forerunners of a Russian force hurrying to their aid. The shooting stopped and, very decently, the Koreans moved such transport as they had out of the way. Perhaps we even waved to each other. Meanwhile I drew my revolver with somewhat of a clammy hand and looked at my watch. I therefore know that it took exactly 17 very long minutes to review the North Korean brigade.'

The Argylls' Mortar Platoon, apparently became even more embroiled with the North Koreans, who definitely thought they were Russians, largely thanks to the woollen cap comforters that they were wearing. Neilson again:

'It was thought as well to maintain the deception for the moment. An entente was established, cigarettes and unintelligible conversation was exchanged, and Lt Fairrie, the platoon commander, had a North Korean "comfort girl" issued to him. She immediately got in his jeep and further cemented relations by exchanging hats with him. It was, however, a situation with no future and it ended, as it had to, with an exchange of shots as a result of which the Koreans made off and Fairrie lost his "comfort girl". He did, however, managed to retrieve his balmoral.'

The advance ended on the Yalu River, but in November the Chinese forces joined in, driving the UN forces back to the 38th Parallel, where the situation was temporarily stabilized. The Chinese resumed their offensive in January, which the UN forces held after giving some ground. They, in turn, launched a counter-offensive and by early April the line was once more stabilized on the 38th Parallel.

During this time a second British brigade, 29th, had arrived in Korea. This consisted of the 1st Royal Northumberland Fusiliers, 1st Glosters and 1st Royal Ulster Rifles. Taking a Belgian battalion under command, the Brigade took up positions on the Imjin River once the line was stabilized. Their task was to cover two critical defiles through the hills to the north. It so happened that the Chinese were preparing yet another offensive and that one of their main axes would come straight through the defile covered by the Glosters, the left-hand forward battalion of the Brigade. The Chinese began their twenty-mile approach march on the evening of 21 April. The following day was a Sunday and was clear and crisp after a heavy overnight frost. The Fusiliers had a church service to mark St George's Day and that night the Glosters set up listening posts down by the river to give early warning of the Chinese approach. One of these, by a ford over the river, which later became known as Gloster Crossing, was manned by three men, Corporal George Cook, Private 'Scouse' Hunter and Drummer Anthony Eagles. They spotted fourteen men approaching the river and informed the Adjutant, Captain (later General Sir) Anthony Farrar-Hockley, who got the gunners to put up flares. He then gave orders that the Chinese must not be allowed to cross. Drummer Eagles:

'When they [Chinese patrol] were in exactly the position, the two of us opened up with rapid fire. We had a rifle, 50 rounds and two grenades each, but so quickly did we let go at them that the adjutant asked Cpl Cook who had the Bren gun! "It's Eagles and Hunter with rifles only, sir," he replied. We succeeded in killing three of them and their bodies floated down the river. We had apparently wounded four others, as their comrades carried them hurriedly back to the north side and into cover. There was no further action by us, but I can assure you that for the rest of the night we were on a knife edge, as we waited for some form of retaliation.'

Other patrols and listening posts engaged the Chinese, but their main effort came against the left-hand forward company, 'A', although they also began to exploit the gap between the Glosters and Fusiliers. 'A' Company, outnumbered by six to one, was soon under intense pressure, and by daylight the Chinese had managed to infiltrate between the forward platoons and had seized a key piece of ground called Castle Hill. The company commander ordered Lieutenant Philip Curtis, who was attached to the Glosters from the Duke of Cornwall's Light Infantry and was commanding the depth platoon, to counter-attack. This he did, but came under heavy fire from a bunker on the top of the hill. Curtis was wounded and dragged back under cover. He refused medical attention, however, and, ordering his men to give covering fire, dashed forward to eliminate the machine gun which was providing all the trouble from the bunker. No sooner had he thrown a grenade into it than he fell, riddled with bullets. It was an act of the highest self-sacrifice and led to the award of the second posthumous Victoria Cross won in Korea. Curtis's company commander was also killed and it was left to CSM Harry Gallagher to withdraw the remnants of the forward platoons to the rear platoon position.

D and B Companies, who were also positioned forward, were heavily embroiled as well, and by midday on the 23rd the Chinese had managed to infiltrate around the rear of the Glosters, thus cutting them off. Luckily, most of the wounded had been evacuated, and food and ammunition delivered before this happened. Lieutenant-Colonel James Carne, commanding the Glosters, now received orders from the brigade commander that he was to hold on at all costs and that efforts would be made to reinforce him on the 24th.

On the night of the 23rd/24th the Chinese, whose lead division had suffered heavily, threw in two other divisions. Desperate fighting took place, with the Glosters inflicting heavy casualties, but further ground had to be given. The rest of the brigade was also under severe pressure and could not help, and a Filipino unit, supported by a tank squadron of the 8th Hussars, spent the whole of the 24th trying to force its way through to the Glosters, without success. That evening Carne pulled in his perimeter, his battalion now down to 50 per cent effectives, to defend Gloster Hill, which still dominated the main route to the south. Attempts to resupply the Glosters by air were largely unsuccessful because of the intensity of the combat and no fighter ground attack could be used. Throughout the night the Chinese attacked relentlessly, until shortly before dawn they had established themselves on the upper slopes of the hill.

'The Chinese trumpets sounded all around ... In retaliation the Adjutant suggested to the Colonel that the Drum Major might give them a reply. This he did and being the man he was, he was not going to do this crouched in a slit trench. Standing up "Drummie" Buss gave the Chinese the full repertoire. Long Reveille; Defaulters; Cookhouse; Officers Dress for Dinner; the Company calls – in fact everything except "Retreat". Buss was a good player and the notes of his bugle echoed over the Glosters' position as the dawn broke.'

With the coming of daylight it was possible to mount some air attacks against the Chinese, who were hit with napalm and machine-gun strafes. A second relief attempt failed before it almost got started and, by mid-morning, it was clear that the Glosters were on their own. James Carne gave orders for his companies to break out independently, but he himself, together with the medical officer and sergeant, and padre, elected to remain behind with the wounded. Of the 622 Glosters who had gone into action, only five officers and forty-one men were left at the end of the battle. Carne was awarded the Victoria Cross for his outstanding leadership and, besides Curtis's VC, no less than twenty-two other decorations were awarded to the Battalion.

Those who were taken prisoner had to endure over two years as guests of the North Koreans, who made every attempt to subdue their spirits and brainwash them, as is graphically and movingly described in Tony Farrar-Hockley's now classic *The Edge of the Sword*. Their conduct in captivity was recognized by thirteen further awards, including a posthumous George Cross to Lieutenant Terence Waters, attached to the Glosters from the West Yorkshires. Further recognition of the Glosters' performance came in the award of the US Presidential Citation, whose streamer was trooped on the pike of the Regimental Colour every 'Black Badge' Day, the annual commemoration of the Glosters' equally famous performance during the Battle of Alexandria in 1801. Every member of the regiment also thereafter wore a symbol of the Presidential Citation in the form of a blue riband with gold border worn on the left upper arm. Finally, the Press gave the Glosters a new nickname, 'The Glorious Glosters'.

Some 30 miles further to the east, 27 Brigade, which was redesignated 28 Commonwealth Brigade at midnight on 25/26 April, was also engaged in a severe fight in the Kapyong area. The Brigade still had the Middlesex, Argylls and 3rd Royal Australian Regiment, although the Argylls were in the process of being relieved by the 1st King's Own Scottish Borderers. In addition there were the 2nd Princess Patricia's Canadian Light Infantry, 16 New Zealand Field Regiment, part of the 5th US Cavalry Regiment and a US tank battalion. They inflicted heavy casualties on the Chinese, forced them to look elsewhere for success, with the Australians and Canadians also being awarded US Presidential Citations. Indeed, although the actions on the Imjin had driven back the UN covering forces, the Chinese were unable to penetrate the main defensive line and eventually they were forced back to the 38th Parallel. After this the British Commonwealth forces were concentrated together under the umbrella of the Commonwealth Division, which came into being on 28 July, 1951.

The last two years of the war were relatively static as the seemingly endless negotiations to bring it to an end dragged on. As such, it took on much of the character of trench warfare, with both sides constructing elaborate defensive systems and trying to combat the intense cold of the Korean winter and the heat of the summer. This is not to say, however, that there were not significant offensive actions. One heavy Chinese attack brought about the award of the fourth and last Victoria Cross of the Korean War, won by Private Bill Speakman of The Black Watch, but attached to The King's Own Scottish Borderers, for his actions on 4 November, 1951, in helping to repel it, which included at one point throwing rocks, ration tins and empty beer bottles at the Chinese because he had run out of grenades.

Another attack occurred only two months before the armistice was signed. This involved the defence of the notorious Hook position, which helped to dominate the route from the north through the Samichon River valley. For some months the Chinese had set their heart on taking it and tried first to wrest it from the US Marines in October, 1952, and then a month later from the 1st Black Watch. In May, 1953, they tried again, first testing the defences, held again by The Black Watch, on the night 7th/8th. There was then a lull, during which time the Black Watch were relieved by the 1st Duke of Wellington's, who now had on their left the Turkish Brigade and on their right 1st King's. The Hook itself was held by one company under Major Lewis Kershaw, and before dark on 28 May the Chinese suddenly opened a heavy mortar and artillery bombardment on the feature. They attacked after just a few minutes, following up closely under the intense barrage, and soon overcame the forward platoon. Further attacks followed, until eventually the survivors of Kershaw's company were driven into the bunkers under the hill. Kershaw himself, after furiously trying to keep the enemy at bay with grenades, was also forced into a bunker with some ten of his men. As he entered it a Chinese grenade wounded him in the leg and he lost his Sten gun. Grabbing another from a sergeant, he opened fire and continued to do so until hit again by another grenade. The Chinese then set light to the bunker and blew in both ends with explosives, trapping Kershaw, whose lower right leg was totally shattered, and his men.

In the meantime, Lieutenant-Colonel Ramsay Bunbury had organized a counter-attack, which, with support from 1st Kings, was eventually successful and by dawn had driven the Chinese off. It was a relieved Lewis Kershaw and party who eventually managed to get out of their subterranean trap, and, although his lower leg was later amputated, he was 'certainly much richer in experience and with a sound confidence in the British private soldier, who is more than a match for any enemy'. He was awarded the DSO, as was his commanding officer, and six other Dukes were also decorated. It was to be the last major British infantry action until the Falklands War. Eventually the armistice was signed on 27 July, 1953, the war having cost the lives of 600 British soldiers, many of them National Servicemen.

The commitments in Malaya and Korea and the sharpening of the Cold War in Europe, especially after the 1948–9 Berlin Airlift, forced the British Government to take urgent steps to strengthen the Armed Forces. The existing six divisions overseas were to be brought up to full strength and the strategic reserve in Britain was to be increased to a full infantry division, the 3rd, which had been disbanded in 1947, and a full armoured division, the 6th, together with 16 Independent Parachute Brigade Group. In order to find the manpower for this, National Service was increased to two years, with a compensatory reduction in reserve service to three and a half years. Additional operational infantry battalions were found by relieving them of the brigade training group role and by allowing seven regiments to reform their 2nd Battalions.

Conscious of the might of the Red Army's arsenal, especially the number of tanks, efforts, too, began to be made to modernize equipment, which was still entirely that left over from the war. For the infantry the first concrete indication of this was when their 6-pdr anti-tank guns were replaced by 17-pdrs. These had previously been the province of the Gunners, who now surrendered the anti-tank role. In turn, the 17-pdr was replaced by the recoilless 120mm BAT (Battalion Anti-Tank gun), while the platoon's PIAT was succeeded by the 3.5-inch Rocket Launcher, and the section received the Energa anti-armour rifle grenade.

Additional defence burdens were not long in coming. In October, 1951, the Egyptian Government decided to tear up the treaty which gave the British the right to use the Canal Zone as a Middle East base and ordered the garrison to leave. This was immediately followed by civil unrest and elements of the 1st Division from Cyprus and Libya and 16 Parachute Brigade in Cyprus were immediately despatched to the Canal Zone. The newly reformed 3rd Division also became involved, with 19 Brigade flying out to Libya to relieve 1 Guards Brigade, earmarked for the Canal Zone, while the remainder of the Division was sent by sea to the Mediterranean. The Royal Navy used its aircraft carriers *Illustrious* and *Triumph* for this. The Border Regiment's journal describes their departure:

'At Portsmouth we were met by the Band and together with the Drums, they marched us to the dockyard where the Commander-in-Chief Plymouth, accompanied by General Sir Gerald Templer and the Divisional Commander, took the salute as we approached HMS *Victory*.

'Preceded by the Colours and Escort, the Battalion boarded *Illustrious* and we were met by the Colonel of the Regiment, who later in the evening came down into the hangar and amongst 2,000 hammocks bid us farewell.

'The hangar, which housed practically everyone for both sleeping and messing, had been converted by the Royal Navy Dockyards in five days. It was an immense undertaking which included welding to the steel floor some 300 army-type trestle tables and 600 forms. In addition, steel hawsers for the slinging of 2,000 hammocks were fixed. Lashed to the flight deck above were 150 assorted vehicles of the Division.

'At 0915 hours on Monday 5 November, HMS *Illustrious* was towed out into the harbour, all troops were assembled on the flight deck, and as we left the quay our band, together with that of the Buffs, played incidental music and the Regimental Marches. At 0930 hours, with the ship under its own power, we all turned to starboard to salute the flag of the Flag Officer Submarines and so slipped out into the English Channel.'

Initially, 3rd Division's role was to prepare for an amphibious landing at Alexandria in order to rescue British nationals, and this very nearly came to pass towards the end of January, 1952, when Cairo erupted into mindless attacks by the mob on anything of European origin. Luckily, King Farouk acted, ordering his army to clear the streets and the tension lessened.

Thereafter 1st and 3rd Divisions were committed to the protection of the Canal. This was a monotonous task, with often 80 per cent of a battalion's strength being committed to guards and few facilities for relaxation. The situation had improved by late 1952 for exercises to take place in the desert, and these certainly did much to relieve the monotony. At the same time it became possible to begin to thin out the troops, but the commitment did not end until June 1956. Yet, as the Commanding Officer of the 1st Beds and Herts told his men at the end of their tour:

'There is no excitement to look back on, of battle in Korea, or jungle fighting in Malaya or Kenya, but rather of arduous and monotonous duties in sand, flies and heat, with living conditions and amenities much to be desired. Such tasks called for greater fortitude, high morale and discipline; these you displayed and maintained to the highest degree throughout.

The increasing use of aircraft in conjunction with ships to transport troops to and from overseas meant that tours abroad were very much shorter than they had been prior to 1939. Even so, as late as 1952 the 1st Royal Fusiliers returned to Britain, having been overseas for thirty years since being posted to India in 1922.

The next trouble spot was Kenya, where an element of the Kikuyu tribe formed a secret society, the Mau Mau, pledged to wresting the fertile White Highlands off the European farmers. In Autumn, 1952, there were murders of Africans loyal to the British administration, and attacks on cattle and buildings on European farms. There were no British troops stationed here, although there were battalions of the King's African Rifles, and in October the Governor felt forced to declare a state of emergency and ask for military assistance.

First to arrive were the 1st Lancashire Fusiliers from the Canal Zone, but this was not sufficient to prevent further attacks on Africans and on Europeans, in spite of a round up of suspected Mau Mau leaders. Consequently, 39 Brigade, recently returned from the Canal Zone to Britain, was sent out. The Mau Mau themselves were ill-equipped, and indeed relied to a large extent on homemade firearms, sometimes more dangerous to the user than to his enemy, and were, unlike the terrorists in Malaya, ill-trained. This initially led to a degree of complacency among the security forces.

The terrain, however, was just as challenging as Malaya. The forests of the Aberdares in which the Mau Mau were based, were dense, largely bamboo up to 11,000 ft and above this was tall jungle grass, and the troops, especially those with no experience of jungle fighting, had to learn from scratch. Mistakes were made, the Commanding Officer of The King's Own Shropshire Light Infantry was shot by his own men when he walked into an ambush which they had laid. Nevertheless, by employing exactly the same strategy as used in Malaya, the Mau Mau were gradually contained, although an additional British brigade of two battalions had to be deployed in Autumn, 1953. The Mau Mau managed to maintain their strength, however, and even increase it, largely through intimidation, but numbers were not enough and by November, 1956, when the last British battalion left, there were well over 30,000 Mau Mau in detention camps and gaols, and a further 11,500 had been killed. As in Malaya, the infantry became highly skilled at operating against men to whom the forest was a second home and there were some notable successes. The Buffs, for instance, who came out with 39 Brigade, killed 290 and captured 194 during their two years in Kenya for the loss of just one man killed.

During the mid-1950s some respite for the stretched British Army was given by the withdrawal of the occupation forces from Austria and the 2nd Lancashire Fusiliers and 1st Loyals from Trieste, as well as the end of the Korean War and withdrawal from the Canal Zone and Sudan, where the 1st Royal Leicesters were the last to be stationed. Indeed, in 1955 it was announced that the reraised 2nd battalions were likely to be disbanded once more within the next two to three years.

Yet fresh trouble spots kept emerging. Thus, the election of Dr Cheddi Jagan's militant left party in British Guiana in 1953 caused the British Government such concern that it felt forced to suspend the constitution. Unrest followed and two companies of the 1st Royal Welch Fusiliers, then based in the Caribbean, had to be sent there. They, in turn, were replaced by the Argylls, and then The Black Watch. In 1956, however, the situation had so improved that the Worcesters, now the resident battalion in Jamaica, were able to keep the peace with just one detached company.

More serious, however, was Cyprus, now a very important strategic base. Here the tensions between the Greek and Turkish communities increased sharply in 1954 when Archbishop Makarios began to champion the island's union with Greece. Two battalions, the 2nd Green Howards and 2nd Royal Inniskilling Fusiliers, had to be sent from the Canal Zone to help contain the growing unrest, but it was not until April, 1955, that matters began to get seriously out of hand. Colonel George Grivas, a retired Greek Army officer, set up a terrorist movement, EOKA, in the Troodos Mountains. They began to set off bombs and then followed this by attacks on police stations. By the autumn additional battalions had had to be deployed and a full-scale counter-insurgency campaign was underway.

This took two distinct forms. First, there was the rural aspect, the tracking down of the EOKA bands in the mountains, but also there was trouble in the urban areas. Rioting, often with Greek schoolchildren in the forefront, and bombings became commonplace. The situation was further compounded by the fact that the two communities continued to be at each other's throats and each believed that the British was favouring the other. As the British Government tried desperately to find a political solution to the problem matters seemed to go from bad to worse, and by mid-1958 there were no less than twenty-six battalions, including Royal Marine Commandos, on the island. Part of this force, it is true, was made up of two brigades of 3rd Divison, which had been hurriedly deployed to Cyprus in July of that year as a precautionary measure to protect British interests in Iraq following the assassination of the King and his prime minister. Nevertheless, during the four months they were on the island they provided a welcome reinforcement, especially since at this time attacks against the British Army peaked. Some of the worst EOKA atrocities were the killing of a sergeant, who was walking with his small son in Ledra Street, Nicosia, the so-called 'Murder Mile', and the shooting of two Army wives, one fatally, while they were shopping in Varosha. The latter incident was one of the few times when the feeling of the soldiers boiled over. One hundred Greek Cypriots were rounded up and taken to detention centres, where some were roughly handled and one died. Grivas himself was never caught, although a number of his chief lieutenants were run to earth, but, eventually, in Summer, 1960, a plan for independence, a mixed Turkish-Greek government, was agreed by all sides and the emergency came to an end. The British were allowed to retain two sovereign bases at Dhekelia and Akrotiri. Cyprus's troubles, however, had not been solved for good.

At the end of July, 1956, a month after British troops had withdrawn from the Canal Zone, and with the counter-insurgency campaigns in Malaya, Kenya and Cyprus still at their height, President Nasser of Egypt delivered a bombshell, declaring that he was nationalizing the Suez Canal. This caught the British and French, the principal shareholders in the Suez Canal Company, by surprise, but the immediate instinct was that force might be required to regain control of the Canal. Conscious that the Egyptian forces had recently been re-equipped with modern weapons by the Soviet Union, the planners believed that a large force would be necessary. Therefore, at the beginning of August no less than 20,000 reservists were recalled to the Colours and 3rd Division was ordered to mobilize complete. This was not easy, since four of its nine infantry battalions were already committed overseas. One of its brigades, 19th, in fact had no infantry, although the 1st Argylls were due to join it from Berlin later in the month. The remainder of the shortfall was made good by bringing in the 1st Royal Scots from Elgin, in the north of Scotland, which meant that the Royal Guard at Ballater had to be found at very short notice by recruits from the Highland Brigade Depot, and the 1st West Yorkshires from Northern Ireland. Within a few weeks the south of England seemed to be awash with military vehicles in sand-coloured paint.

Yet political controversy, both at home and abroad, and the planning complications of mounting a major expedition to a 'far distant shore', which were aggravated by the need to produce a joint plan with the French, meant that it was not until the end of October that Operation MUSKET-EER was finally mounted. With the Egyptians diverted by the Israeli thrust into Sinai, 16 Parachute and 3 Commando Brigades, together with their French allies, had little difficulty in securing their immediate objectives, Ports Said and Fuad at the northern entrance of the Canal before accepting a UN-imposed ceasefire. It was now that perhaps the trickiest part, from the military aspect at least, of the whole affair came. The UN intention was to send a force to secure the Canal and prevent further hostilities, but it would take time to organize this. In the meantime the forces on the ground had to watch and wait.

A week after the ceasefire two brigades of 3rd Division, which had left England in troopships on 1 November, landed at Port Said and relieved the Paras and Commandos. While 19 Brigade looked after Port Said, 29 Brigade covered the causeway stretching 20 miles south to El Cap, where the 1st Royal West Kents sat opposite the Egyptian Army. The latter, however, was not really a threat and strictly observed the ceasefire. The trouble came from what was called the 'Black Hand', urban guerrillas organized by the Egyptian Police Special Branch. Major (later Major General) Eddie Fursdon was on the staff of 19 Brigade:

In a remarkably short space of time our soldiers adjusted themselves to a life of constant patrol and guards. They learnt to be suspicious, cautious and extremely observant. I shall never forget watching one Royal Scots patrol working its way slowly along the wide street that separates Arab Town from Shanty Town. It was a patrol of seven, operating in the rough shape of a diamond. The rear Jock walked slowly backwards, his loaded bren gun, slung from his shoulder, was held at the hip; slowly and continuously the barrel traversed the street, but I am sure that it was his expression that made for the perfect behaviour of the crowd. His bonnet was back a little on his head; his young sunburnt face was set; it was quite clear to everyone, "There will be no nonsense here". There wasn't.

'The success of the operation was largely due to the high standard of junior leadership. It was the corporal, sometimes just a private soldier, who led the patrol and got to know his area as well as the village policeman does at home. A grenade on a cafe table was seen by a passing Royal Scot. The printing press that produced most of the anti-British leaflets was discovered down a side street by Lance Corporal Furness of the West Yorkshires, all because he spotted that only two out of four shops' shutters were padlocked. Corporal Armour of the Argylls became suspicious of two Arabs carrying baskets from a rowing boat. He gave chase and, as a result, unearthed a large dump of ammunition and explosives smuggled from the Nile Delta into the fishing village of El Qabuti.'

Towards the end of November the UN troops began to arrive and within two weeks sufficient had arrived for the bulk of the British troops to depart. This left just 19 Brigade, with some supporting elements, in Port Said and it was now that the guerrillas began to strike. On 10 December Second Lieutenant Tony Moorhouse of the West Yorkshires, who the previous day had caught seven terrorists in a house, was kidnapped from his jeep while attempting singlehandedly to arrest a youth posting an anti-British sticker. His disappearance became a *cause célèbre*, but, although his belongings became a centrepiece of an anti-Western museum set up by Nasser, his body was never found. Grenade and sniping attacks rapidly increased, rising to a crescendo on 15 December when a Royal Scots company commander was killed. That night the Royal Scots, supported by two tanks, went into Arab Town and gave the Black Hand such a lesson that the last few days of 19 Brigade's stay passed almost without incident.

Suez cost Anthony Eden the premiership, and Harold Macmillan, his successor, came into office determined to overhaul Britain's defence policy. Large overseas bases were to be run down and locally raised colonial forces increased. Any need for reinforcement overseas would in future be done by airlift from Britain. Furthermore, the advent of the tactical nuclear weapon meant that any future war in Europe was likely to be short. Hence the need of large reserve forces was no longer necessary. Thus, Duncan Sandys, in his famous 1957 Defence White Paper, was able to announce the abolition of National Service in 1960, although the last conscript would not actually complete his service until May, 1963, and reduction in the Army's standing strength from 375,000 to a target of 165,000, all volunteers, by the end of 1962. This would inevitably hit the Infantry hard and there was a very real possibility that the traditional regimental system would vanish altogether.

CHAPTER TWELVE

Farewell East of Suez
1960–1968

THE concept of the Regional Brigade Groups had continued to create controversy throughout the 1950s. Some infantry officers believed that the regimental system was outdated and that tradition was the principal shackle that held the infantry back from being truly effective in the modern postwar world, although when stating this in public they were careful to write under pseudonyms. Others were vehement in its defence. Lieutenant-Colonel (later Brigadier Sir) Bernard Fergusson of Chindit fame perhaps summed up their views in the concluding paragraph of an article he wrote in 1950, when he was commanding the 1st Black Watch:

'We want infantry which will be happy and contented in peace, and which will go to war under the best auspices with the highest possible self-confidence and esprit de corps. It must consist of officers and men with no feeling that they can be posted here and there as mere "bodies" with orders and numbers attached. They must have the feeling that their pride in their regiment is sympathized with and supported. They must be able to draw inspiration from the continuity of their regiment with their countryside or family or both, and have a close affinity with the men who make up their Territorial battalions. Lastly, they must realize if, in the exigencies of war, "grouping" comes, it will only be because it is unavoidable as a war-time measure.'

Now, with the end of National Service on the horizon and the Army about to suffer its most swingeing 'peacetime' cuts, the threat to the Infantry's regimental system and its replacement by a Corps of Infantry loomed even greater.

The Army Council's view was that the reshaped Infantry must be based on the existing Regional Brigade Group system and that these should be converted into 'large regiments'. As a result, apart from immediately disbanding the seven reraised 2nd battalions,* the plan announced in 1957 was that the regiments within each Regional Brigade Group would now all wear the same brigade capbadge and that there would be one brigade depot. While each regiment was permitted to retain its drums, pipes and bugles, there would again be just one brigade band. The following year there were some adjustments within the brigades as the next step towards creating large regiments. The three English Fusilier regiments were transferred to the new Fusilier Brigade and the titles of the Yorkshire and Northumberland Brigade and Midland Brigade were changed to the Yorkshire and Forester Brigades. Consequently by Autumn, 1958, the brigades were as follows:

Brigade	Depot
Guards	Caterham
Lowland	Midlothian
Home Counties	Canterbury
Lancastrian	Preston
Yorkshire	Strensall
Forester	Leicester
East Anglian	Bury St Edmunds
Wessex	Exeter
Light Infantry	Shrewsbury
Mercian	Lichfield
Welsh	Crickhowell
North Irish	Ballymena
Highland	Aberdeen
Green Jacket	Winchester
Gurkha	Singapore
Fusilier	Sutton Coldfield

In addition, in 1958 the Parachute Regiment, whose officers had until then been seconded from other regiments, was given a permanent officer cadre for the first time, and its depot was confirmed as Aldershot.

Rather than disband individual regiments, a policy of amalgamation was adopted. There was, understandably, intense lobbying to preserve individual regiments, but the Army Council stuck to the principle of the junior regiments in each brigade being amalgamated, although, in some cases, this was tempered by territorial proximity. The first to take place was on 25 April, 1958, when the East and West Yorkshires combined to form The

---

* These came from the Green Howards, Lancashire Fusiliers, Royal Welch Fusiliers, Royal Inniskilling Fusiliers, Black Watch, Sherwood Foresters and Durham Light Infantry.

Prince of Wales's Own Regiment of Yorkshire and the last was on 1 March, 1961, when The Buffs and Royal West Kents became The Queen's Own Buffs, The Royal Kent Regiment. The result, in terms of Regular battalions, was as follows:

Brigade	Pre-1958	Post-1961	
Guards	10	8	Grenadiers and Coldstream lost their 3rd Bns
Lowland	5	4	
Home Counties	6	4	
Lancastrian	7	4	
Fusilier	3	4	Reflecting 1962 granting of Fusilier title to Royal Warwicks
Forester	3	2	ditto
East Anglian	6	3	
Wessex	6	4	
Light Infantry	5	4	
Yorkshire	5	4	
Mercian	4	3	
Welsh	3	3	
North Irish	3	3	
Highland	5	4	
Green Jackets	3	3	
Parachute Regt	3	3	
Gurkhas	8	8	
Totals	85	68	

Where the Army Council did tread more carefully was over the large regiments. Two, however, were formed from the brigades during this period, their members believing that this would help to safeguard their identity in the future. In November, 1958, The Green Jackets came into being, its three battalions being made up of the Oxfordshire and Buckinghamshire Light Infantry (1st), 60th Rifles (2nd) and Rifle Brigade (3rd). The connections among the three regiments had always been strong and went back to the famous Light Division of the Peninsular War. Also, the battalions of the new regiment retained some of their individual idiosyncracies of dress, including the 43rd and 52nd officers' unique tradition of wearing a white tie with mess dress. Thus, this proved to be a relatively straightforward exercise. The other new regiment, the East Anglian Regiment, came about largely because the East Anglian Brigade was the only one to suffer amalgamation of all its regiments and there were problems in trying to agree new titles for them. This was especially in the case of the Beds and Herts and the Essex Regiment, who formed the 3rd Battalion, the other two battalions being the 1st (Royal Norfolk and Suffolk) and the 2nd (Duchess of Gloucester's Own Royal Lincolnshire and Northampton). Unlike the Green Jackets, the only distinction in dress among the three battalions was the colour of the lanyard.

While there was sorrow at the disappearance of famous names from the infantry's order of battle, the amalgamations worked surprisingly well, not the least reason being that many of the new regiments quickly found themselves embroiled in the British Army's continuing overseas commitments. Malaya and Cyprus continued well into the amalgamation programme, but new crises soon developed.

As for recruiting for the new all-Regular Army, this proved to be disappointing, in spite of attractive new rates of pay and a three-year short-service enlistment, which had been introduced in 1952. Part of the problem was the generally healthy economic situation, but also that the 165,000-man ceiling that the Government had set was not sufficient to meet the Army's commitments; hence a revised target of 180,000 was agreed. In order to help make good the immediate shortfall, over 9,000 of the last batches of National Servicemen had to serve an additional six months, which explains why the last of them did not leave until Spring, 1963.

The end of the Fifties saw the Infantry receive more of the fruits of the procurement of new weapons which had been set in train at the beginning of the decade. First and foremost came a new family of small arms. The more robust and better engineered Sterling sub-machine gun took over from the Sten. More important, a new rifle was brought into service. There had been a rifle developed by the Royal Small Arms Factory at Enfield in the late Forties and early Fifties. This had drawn on the lessons learnt during 1939–45, which above all called for a lighter weapon with a much higher rate of fire than the SMLE No 4. The solution was to adopt a smaller calibre of ammunition, 0.280-inch (7mm), and to give the rifle a 20-round magazine, as well as making it self-loading and thus semi-automatic. It was 9 inches shorter than the No 4 and some 11 ounces (300 grams) lighter. Although it performed extremely well on trials, Britain's NATO allies could not be persuaded to adopt this new calibre and the No 4 continued in service. By the mid-1950s, though, it was clear that the European allies were favouring 7.62mm and the Belgian *Fusil Automatique Léger* (FAL), or FN after the *Fabrique Nationale d'Armes de Guerre* who produced it, seemed to meet the British requirement best. A licence to manufacture a British version was agreed and this became known as the

Self-Loading Rifle (SLR). Likewise, the Belgians also manufactured a machine gun firing the same ammunition. This was equipped with a bipod as a light machine gun, but a tripod was available to enable it to carry out the medium machine function as well. Again, Enfield began to produce this under licence as the General Purpose Machine Gun (GPMG), replacing both the Bren and the Vickers.

In another effort towards greater standardization within NATO, the 3-inch mortar was succeeded by the more accurate 81mm, developed by the Royal Armament Research and Development Establishment (RARDE). Likewise, the more compact Swedish-designed 84mm Carl Gustav began to supersede the 3.5-inch Rocket Launcher. In addition, new more flexible webbing equipment came into service. Finally, and as a reflection that the Army was once more an all-volunteer one, battle dress was replaced by the smarter and more comfortable service dress for wear when not in the field, where a new green combat uniform was to be worn.

The first fresh crisis of the new decade arose in 1961. In June, and not for the last time, Iraq claimed oil-rich Kuwait for its own. Kuwait appealed to Britain and the United Nations, and within three weeks British troops had been deployed there. First to arrive were 42 RM Commando, who happened to be en route for the Far East in the carrier *Bulwark*. They were quickly followed by 45 RM Commando from Aden and two companies of the 2nd Coldstream, sent from Bahrain. A tank squadron of the 3rd Carabiniers, some of their tanks already in the Gulf on a tank landing ship and the others stockpiled in Kuwait, was also quickly on the scene, with the remainder of the regiment following a little later. Also arriving within days were the 2nd Battalion The Parachute Regiment from Cyprus, 1st Royal Inniskilling Fusiliers, and the recently amalgamated 1st King's (Manchester and Liverpool) from Kenya, together with supporting arms and services. The build-up was achieved with impressive speed and undoubtedly deterred the Iraqis from using force to achieve their ambition. The main problem, however, was the heat, as would again be recognized 30 years later. With temperatures rising to as much as 140 in the shade, a man needed as much as two gallons of water per day to survive and it was only possible to carry out meaningful training during the early morning. Nonetheless the force stuck it out until October, when it was relieved by troops of the Arab League, of which Kuwait was a member.

A more unusual task was given to the 1st Grenadiers. They were sent out to the Cameroons in May, 1961, to help supervise the last six months of the UN mandate before the country became a republic. For transport they recruited the local Fulani ponies, but, like their forbears during the 1922 Chanak crisis, did not forget to put on ceremonial parade for The Sovereign's Official Birthday.

The following year there was trouble in the Far East. In early December, 1962, there was a revolt in Brunei designed to overthrow the Sultan and install a government sympathetic to Indonesia, whose President Sukarno was bent on bringing the whole of Malaya, Singapore and the British colonies in northern Borneo under his sway. After an appeal for help from the Sultan, HQ 17th Division, based at Seremban, reacted with speed. The Queen's Own Highlanders (Cameron and Seaforth), 1st/2nd Gurkhas and 42 RM Commando were quickly deployed and crushed the revolt within a week, the rebels fleeing into the hinterland. The task of tracking them down was given to General (later Sir) Walter Walker, probably the most experienced practitioner of jungle warfare of the day.

The Brunei Revolt turned out to be merely an overture, however. Sukarno's chief concern was to prevent the Malay states, Sarawak, North Borneo and Singapore from uniting to form Malaysia, since this would make his task of creating an empire that much more difficult. Consequently his rhetoric against the concept became more strident and he began to infiltrate parties across the border from Kalimantan (Indonesian Borneo) into Sarawak and North Borneo. When Malaysia came into being in September, 1963, with only Brunei choosing not to join, Sukarno promptly broke off diplomatic relations and a mob sacked the British Embassy in Jakarta.

Then, on 28 September, came the first real clash of the Borneo Confrontation when a group of some 200 Indonesians attacked a small outpost, held by thirty Gurkhas, policemen, and indigenous Border Scouts, and overran it after a fierce fight. The 1st/2nd Gurkhas were deployed by helicopter to set up ambushes on the likely Indonesian routes back to the border and succeeded in accounting for a significant number of the intruders. Further successful actions followed, and after the Royal Leicesters had attacked a camp astride the Sarawak-Sabah border, killed seven and captured large quantities of arms and ammuniton in January, 1964, Sukarno began to talk peace. Negotiations quickly broke down, however, and hostilities resumed in March, this time with Indonesian Regular forces being employed rather than volunteers. This made life more difficult, and the fact that Indonesian groups made landings on the Malaya coast, or were dropped by parachute, meant that some reinforcements for Borneo were kept tied down in Malaya.

Walter Walker, however, now decided to take the initiative. With the approval of the British Government, although it was kept highly secret at the time, he launched Operation CLARET. This entailed sending groups of up to company strength either to attack camps previously reconnoitred by the SAS, on the Indonesian side of the border, or to lay ambushes on likely approach routes to it. The troops involved had, however, to remove all means of identification before they set out and were warned that there was no question of wounded or sick being evacuated by helicopter, since this would make it obvious to the Indonesians what was happening. On the home side of the border a series of company bases were established, normally on the tops of hills. They had elaborate bunkers and trench systems, were wired in, with a profusion of Claymore mines laid just outside the perimeter. Each also had a 105mm pack howitzer, an 81mm mortar, and, the last time that they were to see action, two Vickers machine guns. Resupply was entirely from the air. From the base patrols were sent out, usually for ten to twelve days, with four days in between in order to recover and prepare for the next outing. After about six weeks a platoon would be sent back to battalion HQ in order to recuperate in more comfort. Battalions themselves did a six months' tour in Borneo and no less than fifteen British battalions served here, while the Gurkha battalions each did two tours.

As for patrolling in the jungle itself, this was always a 'strain', as Colonel Mike Dewar, who was a platoon commander in the 3rd Royal Green Jackets ('Royal' prefix granted on 1 January, 1966), recalls:

'There was always the likelihood of an encounter with the enemy, though when it happened it was often unplanned, unexpected and fleeting in nature. . . . The ability to shoot quickly and accurately was undoubtedly the most important requirement in the jungle, a requirement made more difficult by the fact that a few minutes or even seconds of action was almost certainly preceded by weeks, even months without contact with the enemy, a situation guaranteed to dull the senses and deaden the mind. The perpetual dilemma of a patrol commander was whether to follow a track and risk walking into an ambush or treading on an anti-personnel mine or to hack his way through nearly impenetrable jungle at the rate of perhaps 200 yards an hour. Inevitably risks had to be taken or the jungle could never have been dominated. . . . It was a game of cat and mouse, of hide and seek and sometimes even of bluff and counter-bluff.

'At five o'clock it started to get dark as the jungle canopy filtered out the light. Patrols broke track and established a bivouac area for the night, movement after dark in the jungle without artificial light being considered impracticable. A circular perimeter was established, sentries posted, and a cold evening meal quietly eaten. For most of the campaign in Borneo it was considered unsafe to cook in the jungle as the resultant aroma could reveal a platoon's position. At six o'clock – on the dot in the monsoon season – it started to rain. Jungle rain takes an hour to come through the trees then it goes on dripping long after

23. Patrol in Borneo, 1966. Much of the area was, however, dense jungle.

24. Northumberland Fusiliers patrol downtown Aden, 1967.

25. Parachute Regiment patrol in Crossmaglen, South Armagh, mid-1970s. This village had the reputation of being the hardest Republican area in the whole of Northern Ireland.

26. Members of the Light Infantry board RAF Wessex helicopters during the airmobile trial held by I (BR) Corps in Germany during 1983-86.

27. The modern British infantryman (1) — members of The Queen's
Own Highlanders (now The Highlanders) armed with the SA-80
personal weapon.

28. The modern British infantryman (2) — A Milan anti-tank guided weapon system firing post.

29. The modern British infantryman (3) — Warrior mechanized infantry combat vehicles of the 1st Battalion Grenadier Guards on exercise in Germany.

the storm itself has stopped. Often it fell steadily for five or six hours seeping through the holes in the soldier's lightweight shelters and turning the jungle floor to mud. Men slept in boots and jungle-green uniforms unshaven and unwashed. The nights were terribly long – a full twelve hours between dusk and dawn. Even the exhaustion caused by the previous day's march could not fill those hours with sleep. Men lay awake thinking of sex or the cold tin of NAAFI beer waiting at the end of the patrol.'

The need for quick and accurate shooting caused many men to use the American 5.65mm Armalite rifle in preference to the SLR since it was smaller, lighter and easier to handle in the close confines of the jungle. Occasionally, the Indonesians would attack the bases themselves. On one occasion B Company of 2 Para, less two platoons, was attacked by a large group, supported by light mortars and rocket launchers. The Indonesians managed to penetrate the perimeter and a desperate close-quarter battle followed, during which at one point the CSM fired a magazine of GPMG at pointblank range at the attackers. After an hour and half the Indonesians withdrew, having killed two Paras and wounded eight others, but had suffered some thirty casualties themselves.

In March, 1966, growing general dissatisfaction within Indonesia forced Sukarno to hand over power to the army, which had become fed up with his ever-leftward-leaning stance. The Indonesian Communist Party was immediately dissolved and General Suharto began to speak of peace. Eventually, on 11 August, 1966, this was signed and the Confrontation came to an end, the last British battalion, The Queen's Own Buffs, having arrived in theatre in mid-July. In all Commonwealth Forces suffered 114 killed and 180 wounded, of which the British battalions' share was sixteen and forty-one, and the Gurkhas forty-three and eighty-seven. Indonesian deaths alone were over 600.

Operationally, apart from Borneo, 1964 proved to be a busy year elsewhere for the British Army. In December, 1963, fighting had broken out among the Greek and Turkish Cypriots over the 1960 constitution, with both Greece and Turkey egging the combatants on from the sidelines. The fighting began to threaten the sovereign base areas, but Archbishop Makarios, at the urging of the British Government, accepted a British truce force. This was initially made up of the two resident battalions, 3rd Green Jackets and The Royal Inniskilling Fusiliers, but elements of 16 Parachute Brigade and a field regiment RA (in the infantry role), quickly joined them from Britain. Major General Mike (later Field Marshal Lord) Carver, then commanding 3rd Division, and part of his headquarters then flew out to take control.

Again the British found themselves unpopular with both sides, but were able to prevent open war from breaking out. In early March, however, the United Nations passed a resolution establishing a UN peacekeeping force on the island. Mike Carver was appointed as deputy commander, with the force being made up of British, Canadian, Danish, Irish, Finnish and Swedish troops. The Inniskilling Fusiliers were deputed to look after the Irish battalion when it arrived. Colonel John Tomes, who was on Mike Carver's staff:

'It was interesting to see units from the North and South alongside each other. It was also surprising to the Irish to find that the CO of the Inniskillings was a Catholic. The two units also discovered quite quickly that they had a number of men in their respective ranks who had deserted from the other side and re-enlisted. However, they got on very well and the officers of the Irish battalion gave a special dinner to which only British officers were invited. This was a thank you for our help in getting them settled in. It must have been the first occasion on which an Irish regiment had drunk the Loyal Toast – we also drank to the President. This then became the occasion for the alternate singing of Loyalist and Republican songs.'

In July Carver and his headquarters returned to England, with the British presence in UNFICYP now reduced, but from then until this day there has always been a British infantry battalion, together with a reconnaissance squadron and logistics elements, wearing the UN's blue beret in Cyprus.

There was trouble, too, in South Arabia. During the late 1950s and early 1960s the British had persuaded most of the twenty sheikdoms in the hinterland around Aden to join together in the Federation of South Arabia in order to improve stability and to enable them to better protect themselves. This was not to the liking of President Nasser of Egypt, nor to Aden's northerly neighbour, Yemen, and leftist elements within Aden itself. While there were one or two incidents within Aden itself, the Egyptians were more concerned to stir up the tribes in the mountainous hinterland, especially those of the Radfan, which lay 60 miles north of the port of Aden and adjacent to the road running from there to the border with Yemen. An operation was therefore mounted in January, 1964, by the British-officered Federal Regular Army, supported by tanks of the 16th/5th Lancers, and British guns, engineers and aircraft. They succeeded in penetrating the Radfan, but did not have the troops to secure it. They were thus forced to withdraw and this merely served to encourage the dissidents, who made numerous attacks on the Dhala road.

At the end of April, therefore, a larger operation was mounted with troops garrisoning Aden. These included 45 RM Commando, the 1st East Anglians, a company of 3 Para from Bahrain, two weak FRA battalions, and supporting arms. The plan was to seize two prominent features in the north initially by an airborne *coup de main*, which would be covered by a diversionary attack up a wadi to their south. The latter was mounted by the East Anglians, supported by armoured cars of the 4th Royal Tank Regiment. This went in first and made steady progress in the face of opposition, but the operation became stalled when the SAS party which was marking the dropping zone (DZ) was spotted and surrounded by dissidents, and had to fight their way out after the patrol commander and one other had been killed. Nevertheless, 45 Commando had established themselves on two other positions to the west of the 3 Para objectives. From here they managed to get to the summit of one of the Para objectives, Cap Badge, but the paras got caught in the open at first light while on the lower slopes of this feature as they were moving to join the Commandos. Consequently it was not until that night that they were able to do so, after a day-long battle.

There was now a pause while reinforcements were deployed to clear the remainder of the Radfan. These included The King's Own Scottish Borderers, the remainder of 3 Para and HQ 39 Brigade under Brigadier (later General Sir Cecil) 'Monkey' Blacker. Operations resumed on 18 May with a spectacular advance by 3 Para for 14 miles along the virtually impassable Bakri Ridge which lay in the centre of the Radfan. Although it was not until 8 June that the Radfan was totally secured, and not before there were some stiff fire fights, 3 Para's thrust was decisive and earned for its commanding officer, Tony Farrar-Hockley, a bar to the DSO which he had won with the Glosters on the Imjin 13 years before.

While the first foray into the Radfan was proceeding trouble flared up elsewhere. In January there was a revolt against the Sultan of Zanzibar. With the troops in Aden already committed, 24 Brigade, still in Kenya, although beginning to run down its strength, was put on alert. First to move were 2nd Scots Guards, who had been deployed to Zanzibar the previous summer in order to provide security during the elections there. In the event they were flown to Aden, and a company of 1st Staffords went by air to Mombasa where they embarked in the frigate HMS *Kelly*. No sooner had they done so than the Sultan fled and they returned to port. Within a few days there was a cry for help from President Nyerere of Tanganyika. There had been a revolt by the two battalions of the Tanganiyka Rifles (formerly King's African Rifles). They had deposed their British officers and NCOs and went on to fly them out of the country to Kenya. The remainder of the Staffords and the Gordons, who were just preparing to leave Kenya for Britain, were again stood to, but 45 RM Commando were quickly sent by sea from Aden and overcame the problem with just one Carl Gustav rocket through the guard room of the 1st Tanganiyka Rifles.

This was not the end, however. On 23 January, three days after Nyerere's *cri de coeur*, Dr Milton Obote requested British help since the epidemic of unrest had spread to the Uganda Rifles (also formerly KAR). The Staffords, less two companies, and Right Flank Company 2nd Scots Guards were therefore flown to Entebbe. On landing the Staffords immediately commandeered vehicles and, leaving the Scots Guards to secure key points in Entebbe, drove through the night 70 miles to Jinja, where the trouble was. They arrived shortly before dawn, catching the Uganda Rifles totally by surprise and quickly putting an end to the revolt, without a shot being fired. There was trouble, too, on 23 January, with Kenyan troops at Lanut, 20 miles from 24 Brigade's main base at Gilgit. This was speedily dealt with by 3rd Regiment Royal Horse Artillery, assisted by Sappers and elements of the Gordons, again without bloodshed. Peace restored, and 24 Brigade could continue with its withdrawal from Kenya, which was completed in October, 1964. The East African mutinies could have had serious consequences for the three new independent states. Swift action, combined with minimum force, ensured that they were snuffed while in their early stages, and without too much rancour. This says much for the troops who took part.

Back across the other side of the Indian Ocean, the Radfan may have been subdued, but trouble was growing in Aden itself. There had been a number of incidents created by the National Liberation Front (NLF), but after the British Government announced its intention in July, 1964, to give South Arabia its independence before the end of 1968, although with the provision that a British base would be maintained at Aden, these began to increase. On 23 December a grenade was thrown into the house of an RAF officer, where a teenage Christmas party was taking place, killing one 16-year-old girl and wounding several others. This marked a stepping-up of the terrorist campaign and was accompanied by an increase in attacks on British forces in the hinterland, 2nd Coldstream suffering three Guardsmen killed in an ambush on the Dhala Road in March, 1965. As in many other postwar counter-insurgency campaigns, the riot was one of the principal means of keeping the security forces tied down.

The drill for coping with riots was by now firmly laid down; the British Army had built up a wealth of experience since 1945. The request for Army assistance would come from the police. The troops would then deploy. If this was not enough to dissolve the crowd, as was often the case, the military commander on the spot, usually a company commander, had to obtain the signature from the government representative present that control was being handed over to the soldiers. This done, and during it the troops were usually being subjected to an onslaught of bricks and other missiles, the soldiers would lay out a line of white minetape between themselves and the rioters and unfurl a banner. This was inscribed with a message in English and the local language warning that anyone crossing it would be shot. Simultaneously, a section of infantry would double round to the front of the sub-unit and adopt the kneeling position. Their weapons were loaded, but usually only one man had live ammunition, the others having merely blank. The message on the banner would be repeated verbally, using a megaphone. If the crowd did cross the line a fire order would be given, the target being someone who the commander believed to be a ringleader, the proverbial 'man in the red shirt' of a training film made on the subject. Almost invariably this was sufficient to cause the crowd to disperse, but a lesson learnt from Amritsar in 1919 was that the crowd must be allowed unhindered exits from the location of the riot. The troops then merely followed up, arresting anyone who seemed unwilling to withdraw. The underlying principle was that of 'minimum force', which often meant the soldier having to bear indignity and injury before he was allowed to retaliate, which, even then, was tightly controlled. It proved a severe test of discipline, but few were the times that the British soldier failed it.

The terrorists' main weapon in Aden was the grenade, although they also began to use rockets and then mortars. In the narrow and often crowded streets it was only too easy to strike at a patrol and then quickly melt away before it could react. In 1964 there had been thirty-six terrorist incidents, but this grew to 286 in 1965, resulting in five British soldiers killed and eighty-three wounded.

The British Government announcement in early 1966 that it now intended to pull everything out of Aden merely served to increase the number of incidents. Worse, it provoked growing disillusion among the pro-British Arabs, who realized that they would have no protection once the British left. Thus, the troops found themselves fighting a rearguard action, which would last until November the following year. During the latter half of 1966 there was a squabble among the various terrorist factions, which temporarily reduced some of the pressure on the troops, but by the end of the year the NLF had come out on top and turned its attention once more to the British. As a result in early 1967 a fourth battalion was sent out, making four in all – 1st Royal Northumberland Fusiliers, 1st South Wales Borderers, 1st Cameronians and 3rd Royal Anglians. The rate of incidents rose rapidly, reaching a climax at the end of April, when a bomb killed nine British children and injured a further fourteen in a school bus. The result was that service families began to be evacuated back to Britain.

On 1 June, 1967, 1 Para, who had relieved the Royal Anglians, went into the notorious Sheikh Othman quarter and, in order to pre-empt a general strike called for that day, occupied a number of observation posts and fought a day-long battle with the terrorists, killing five and capturing a further six at a cost of one killed and four wounded.

Three weeks later, on the 20th, came one of the worst days in Aden. A mutiny, largely caused by misunderstandings, in the newly formed South Arabian Army resulted in its men opening fire on a lorry carrying British troops back from the rifle range. Eight of the occupants were killed, together with two policemen, a British civilian employee and a Lancashire Fusiliers subaltern, all of whom were caught in the cross-fire, and eight wounded. In a remarkably skilful operation a company of The King's Own Scottish Borderers managed to secure the South Arabian Army camp without firing a shot, although one Jock was killed and eight wounded in the process. Sadly, however, the same day saw a Northumberland Fusilier company commander, his CSM and four men, together with a company commander and two men of the Argylls, who were in the process of taking over from the Fifth Fusiliers, caught in an ambush in the Crater area. The company commander had gone to check one of his platoons with whom he had lost radio contact and the two landrovers in which the party was travelling came under fire from the Police Barracks in the Crater, whose inmates had decided to join their army comrades, and all bar one were killed. The Crater was therefore ringed with troops, who fired on anyone carrying an illegal weapon.

Two weeks later, on the night 3/4 July, the Argylls, under Lieutenant-Colonel Colin Mitchell, who became known as 'Mad Mitch' to the media, went in to secure the Crater. It was a superbly executed operation, totally exceeding the original aim laid down by the military commander, Major General Philip Tower, who merely wanted a long slow 'nibbling' operation. Thus, with the two main trouble spots no longer 'no go' areas, the troops could console themselves that their seemingly thankless task had finally put them on top.

Gradually, from the end of August, the battalions began to withdraw, handing over to the South Arabian Army, which was soon to side with the NLF. 1 Para withdrew from Sheikh Othman, which they had held for four months, suffering twenty-four casualties, but killing thirty-two terrorists and wounding thirteen. The South Wales Borderers and Lancashire Fusiliers left the colony, and 42 Commando landed to act as the final rearguard. Now the Argylls left, their hold on the Crater firm to the last, and eventually the troops were pulled back to Khormaksar, where 1 Para, 45 RM Commando and finally 42 RM Commando were lifted out by helicopter to the carrier *Albion*, the last company leaving on 29 November.

Aden and the Radfan had cost the British Army (including Royal Marine Commandos) ninety killed and 510 wounded. Much of the time the soldiers had to perform in the full knowledge that they were merely 'holding the ring' until a political solution could be found. It was not the first time that this had happened, and it would not be the last, but the infantryman, in particular, bore the brunt and did so with remarkable stoicism.

The 1966 decision to withdraw from Aden was part of the Labour defence minister Denis Healey's policy of withdrawing from major commitments east of Suez. Instead, priority would now be given to NATO in Europe. Here the infantryman had been facing a different form of challenge. The main operational implication of the introduction of the tactical nuclear weapon, through the successful US test firing in 1953 of a low-yield nuclear shell from a 280mm gun, was that forces would have to be very much more widely dispersed than hitherto, but would need to be highly mobile in order to be able to concentrate to resist the massive conventional Soviet ground blow that was expected to follow a nuclear strike. While other NATO armies were quick to develop the armoured personnel carrier (APC), or at least rediscover a vehicle that had become increasingly seen on the battlefield during 1939–45, the British were slow off the mark, mainly because the infantry's priority had been on the campaigns that it was waging in the Far and Near East. Although, in 1958, a decision had been made to incorporate an armoured regiment into each infantry brigade, the infantry themselves were still lorried, even in armoured brigades. This was clearly unsatisfactory and in order to improve mobility the 6-wheeled Saracen APC was deployed with 7 and 20 Armoured Brigades in Germany in order to transport the infantry battalion in each brigade. The infantry themselves, however, were mere passengers, since the vehicles were driven and commanded by members of 4th Royal Tank Regiment in the initial belief that the infantry lacked the technical expertise to be able to look after their APCs. This, however, was proved wrong by the 1st Royal Ulster Rifles, who, with the assistance of 1st Royal Tank Regiment, conducted a trial to prove that they were perfectly capable of looking after their own vehicles.

The Saracen, being wheeled, had a serious limitation in that it often could not go where the tanks did, which meant that infantry-tank co-operation suffered. The solution came in 1963 with the introduction into service of the FV432 tracked APC, and soon all the battalions in Germany were equipped with this. Remarkably, it is still in service, 30 years later. Unlike in the recreated German Army, which brought back the Panzer Grenadier concept, the FV432 was no more than a battlefield 'taxi' and lacked the weaponry to be fought from. This was to prove the subject of much debate over the next ten years.

Dennis Healey's dramatic change in the direction of British defence policy inevitably meant that there would be further reductions in the Army's strength. This implied the loss of more Regular infantry battalions. Infantrymen themselves were, however, coming increasingly to believe that their arm needed a radical restructuring. The nub of the problem was set out by Lieutenant-Colonel Colin Mitchell of the Argylls in an article in the *Journal of the Royal United Services Institute* in 1966. The infantry's roles since 1945 had now become too diversified – mechanized in Germany, 'anti-guerrilla' in the Far East, and air-portable as part of the strategic reserve, as well as training of recruits and ceremonial. These each required different organizations, weapons and equipment. The idea that a battalion could switch immediately from armoured warfare to 'jungle-bashing' and maintain the same high standards was unrealistic. The time had come, he believed, to give infantry battalions permanent specialized roles, with a man enlisting specifically to be an armoured infantryman in Germany or to become a counter-insurgency expert. Others, though, were prepared to go further, and there was a school which began to advocate a single combat arm, made up of armour and infantry.

In the meantime the 'large regiment' concept had slowly been inching forward. In 1963, with transfer of the Royal Lincolns to the East Anglian Brigade in 1958 and that of the Royal Warwicks to the Fusilier Brigade in 1962, the Forester Brigade, now only two Regular battalions strong, was broken up. The Sherwood Foresters joined the Mercian Brigade, while the Royal Leicesters were made the 4th Battalion of what was now The Royal Anglian Regiment, a change of title which meant yet another new cap badge for the battalions within it. At the end of 1966 another new large regiment was formed, The Queen's Regiment, from the four regiments of the Home Counties Brigade. Two years later, in 1968, a further three formed. These were the four-battalion Royal Regiment of Fusiliers from the Fusilier Brigade, the three-battalion Royal Irish Rangers from the North Irish Brigade, and The Light Infantry of four Regular battalions from the Light Infantry Brigade.

Almost as soon as the additional large regiments had come into being they faced cuts. Each lost one battalion, the junior, which meant that, to all intents and purposes, all trace of The Lancashire Fusiliers, Royal Irish Fusiliers and Durham Light Infantry was lost from the order of battle. Likewise the 4th Royal Anglians (Royal Leicesters) were disbanded.

Outside the large regiments, others, too, faced amalgamation. Once again the policy was applied that the junior regiments in each brigade would suffer. In two cases, though, regiments elected disbandment rather than amalgamation. The York and Lancaster Regiment in the Yorkshire Brigade was one, and the other was The Cameronians. The latter case was not surprising in view of the unique historic character of the Regiment, which would have made it an awkward bedfellow for the next junior regiment in the Lowland Brigade, The King's Own Scottish Borderers. As it was, the Cameronians maintained their traditions to the last. Their disbandment parade was held in May, 1968, in the form of a Conventicle on the very same spot that the youthful Earl of Angus had raised them 279 years earlier to the day. At it the Reverend Dr Donald McDonald, a distinguished former regimental padre, spoke these words:

'So pride in your step, Cameronians! As you march out of the Army List, you are marching into history, and from your proud place there, no man can remove your name, and no man can snatch a rose from the chaplet of your honour. Be of good courage therefore! The Lord your God is with you wherever you go, and to His gracious mercy and protection I now commit you.'

All in all, this latest round of cuts resulted in the loss of 2nd Scots Guards, ten Line battalions and three Gurkha battalions. Two other regiments, the Royal Hampshires and the Argylls, managed to survive only by the skin of their teeth, largely thanks to intensive lobbying, especially in the case of the Argylls, but both were reduced to company strength. By the end of 1970, therefore, there were just five infantry regiments – Royal Scots, Green Howards, Cheshires, Royal Welch Fusiliers and King's Own Scottish Borderers – who could claim an entirely 'pure' lineage stretching back to their original founding, and only six Cardwell creations which had not suffered amalgamation since, including the Royal Hampshires and Argylls. Nonetheless the Army Board, as the Army Council had now become, had resisted the adoption of a more extreme restructuring, such as a corps of infantry, and had retained the regimental system, even to the extent of not making small regiments wear brigade cap badges.

It was clear, though, that the Regional Brigade system was not too elaborate to administer the Infantry. Consequently, it was replaced, on 1 July, 1968, by a new divisional structure. At the end of 1970 this was organized as follows:

Division	Formed from (Brigades)	No of Regular Bns	Depots
Household	Guards	7	Pirbright
Scottish	Lowland	6 + 1 coy	Milton Bridge
	Highland		(Midlothian)
			Bridge of Don
			(Aberdeen)

The Queen's	Fusilier	9	Royston
	Home Counties		
	East Anglian		
The King's	North Irish	8	Ballymena
	Yorkshire		Strensall
	Lancastrian		Preston
The Prince of	Mercian	8 + 1 coy	Lichfield
Wales's	Welsh		Crickhowell
	Wessex		Lichfield
Light	Light Infantry	6	Winchester
	Royal Green		
	Jackets		

In addition the three battalions of the Parachute Regiment retained their Airborne Brigade Depot at Aldershot, while the Gurkhas moved their training depot from Malaya to Hong Kong in 1971.

During the period 1957–67 there had also been a major reorganization of the Army's reserve forces. There were a number of reasons for this. For a start, the system of having two separate reserves, the Army Emergency Reserve, which supplied individual reinforcements, and the Territorial Army, largely organized in operational formations, was increasingly seen as being somewhat cumbersome. Secondly, the belief that nuclear war in Europe would be of short duration meant that by the time the TA divisions had been mobilized and sent across the Channel hostilities would be over. In any event, the TA was still wholly reliant on Second World War weapons and equipment, and there was not the money available to modernize it. Finally, much of the old volunteer spirit of the Territorials had been heavily diluted by the TA's responsibility for National Servicemen undergoing their reserve commitment.

The first significant step was taken in 1960, when, with National Service now coming to an end, the TA divisions began to be broken up and their headquarters merged with district headquarters, the result being that the latter would now administer not only the Regular units within their geographic area, but the TA units as well. At the same time, the TA was significantly reduced in strength, losing eighteen infantry battalions, among other units. This, however, represented only an initial phase of the main and highly radical reorganization which took place in 1967. The AER and TA were merged into the Territorial Army and Volunteer Reserve (TAVR), which was split into four sections – TAVR I–IV – with varying commitments and terms of service. As for the TA units themselves, the principle adopted was that those which would be too expensive to modernize were to be disbanded. It was the Yeomanry and Gunners who mostly fell victim to this, but the infantry suffered just as drastically, indeed even more so. All the existing TA battalions were disbanded and in their place thirty-eight battalions were formed, almost all with different titles from their predecessors. In most cases these new titles represented part of the new large regiments or anticipated large regiments which were not actually formed. Thus the new order of battle was as follows:

Regiment	No of Volunteer Bns
52nd Lowland Volunteers	2 (1st, 2nd)
51st Highland Volunteers	3 (1st–3rd)
The Queen's Regiment	2 (5th, 6th/7th)
Royal Regiment of Fusiliers	2 (5th, 6th)
Royal Anglian Regiment	3 (5th, 6th, 7th)
Yorkshire Volunteers	3 (1st–3rd)
Royal Irish Rangers	2 (4th, 5th)
Wessex Regiment	2 (1st, 2nd)
Mercian Volunteers	2 (1st, 2nd)
Light Infantry	2 (5th–7th)
Royal Green Jackets	1 (4th)

There was an enormous outcry from the TA county associations and other interested groups, especially in those areas where toally new titles bearing no relation to those of Regular counterparts had been introduced, but only in certain cases was there any redress. The three surviving Welsh TA battalions successfully avoided becoming the Welsh Volunteers, retaining the titles 3rd Royal Welch Fusiliers, and 3rd and 4th Royal Regiment of Wales (formerly the South Wales Borderers and Welch Regiment). Likewise, the two battalions allowed to the north-west of England were titled 4th King's Own Royal Border Regiment (the 1959 amalgamation of the Border Regiment and the King's Own Royal Regiment) and the 5th/8th Queen's Royal Lancashires (formerly The Lancashire Regiment (The Prince of Wales's Own) and The Loyals). The final battalion to escape becoming part of a large regiment was the 3rd Battalion The Worcestershire and Sherwood Foresters Regiment, although logically it should have become part of the Mercian Volunteers. The Parachute Regiment's TA element was also able to avoid these drastic reforms, with three battalions being retained (4th, 10th, and 15th (Scottish)).

The effect on the Territorials themselves was bad. Morale plummetted and recruiting figures fell sharply. Indeed, many looking back on this time afterwards viewed it as the nadir in the history of the TA. In time, though, the situation did improve as the new battalions established their identity. It was helped by the introduction of new weapons and equipment. Also a number of battalions were given the operational role of reinforcing I (BR) Corps in Germany in time of extreme East-West tension, which gave them the opportunity of taking part in large-scale manoeuvres over there.

It was as well that 1968, the height of the turmoil for both Regular and Territorial Armies, was a singular year in the history of the British Army. It was the only year, possibly since 1660, that a British soldier was not killed on operations somewhere in the world. True, there was the UN Commitment in Cyprus and states of emergency in Bermuda and Mauritius, both of which required small groups of troops to be temporarily deployed, but in none of these cases was a shot fired in anger. Matters, however, quickly changed. This time the crisis was on Britain's doorstep.

CHAPTER THIRTEEN

Further Challenges
1969–1982

THERE had been an IRA campaign in Northern ireland during the 1950s. It had begun with raids on various police and army barracks in order to obtain arms. Some, including an attack on The Royal Irish Fusiliers depot at Gough Barracks, Armagh, in June, 1954, were successful, but others total failures. Then, in December, 1956, the IRA began its military campaign proper. The object was to attack the Royal Ulster Constabulary and its part-time B Specials in the hope that this would draw the British Army into the conflict and create an atmosphere of British repression. During the next five years there were a number of incidents, resulting in the deaths of six policemen, but the IRA had nine killed and forty-six jailed. The peacetime garrison of Northern Ireland was built round 39 Brigade, with three infantry battalions and an armoured car regiment, but it took very much a back seat, although it did assist the police with deterrent patrols. Lacking popular support, and with only limited weapons, the IRA called off its campaign in February, 1962, and peace returned to the province. This, however, was not to last long.

The mid-1960s saw the rise of the Civil Rights movement, designed to right the inequalities suffered by Roman Catholics in Northern Ireland. Tension grew and the climax came in August, 1969, in Londonderry during the Apprentice Boys' March to celebrate the 280th anniversay of the lifting of the siege of Londonderry. Protestants provoked Catholics and rioting ensued. The RUC, hopelessly over-stretched, even after they had been reinforced by the B Specials, tried to contain the violence for forty-eight hours, but without success. Thereupon, in the late afternoon of 14 August, the troops, in the shape of the 1st Prince of Wales's Own Regiment of Yorkshire, were called in. Little did they, or anyone else, for that matter, believe that for the next 25 years the violence would remain in being.

To chart the course of the Troubles in Ulster over the past three decades would be to fill a book on its own. Suffice to say that initially the troops in both Belfast, where internecine conflict and police over-reaction resulted in ten killed and 100 injured and the Army being deployed there on 15

August, and Londonderry, were welcomed by the Catholics, who saw them as protection against the Protestant vigilantes. The appearance of the IRA on the scene in early 1970 led to the 'shooting war', when the terrorists engaged the Army on the streets. It was not until February, 1971, that the first British soldier was killed, Gunner Robert Curtis of 94 Locating Regiment RA, and during that year a total of thirty-three soldiers lost their lives in the province. The following year this rose to 103, the peak year for British Army casualties. Then came the period of internment of suspected terrorists, which lasted from 1972–75. Thereafter, from 1977 onwards has been the era in which the RUC took over the lead role, with the Army in support.

The first effect of the Troubles in Ulster was a steady reinforcement of troops deployed there. The overall strength rose from 3,200 in August, 1969, to over 6,000 that October, and thereafter to 11,600 by the end of 1971. Strength peaked temporarily during Operation MOTORMAN, the reoccupation of the Catholic 'no go' areas in Londonderry and Belfast at the end of July, 1972, to 25,000 men. Thereafter the total number of British troops in Ulster has seldom dropped below 10,000. Inevitably, the brunt has fallen on the Infantry, and an immediate result of the demands of Northern Ireland was to halt the reduction in its overall strength. Both The Royal Hampshires and Argylls were brought back to full battalion strength, and 2nd Scots Guards and 4th Queens were reraised at the beginning of 1971, although the last was disbanded again at the end of the following year.

At the end of 1969 the British Government decided to disband the B Specials on the grounds that their attitude was too sectarian. In their place came, in 1970, the Ulster Defence Regiment, initially of seven battalions (one for each Ulster county and the seventh for Belfast), with a further four being raised in January, 1972. It was hoped that this new force would cut across sectarian lines and appeal to both Protestant and Catholic. Initially it did, with the latter forming some 20 per cent of the total strength, but gradually, thanks to intimidation and other reasons, the proportion of Catholics fell. Most of its members were part-timers, but each battalion had a small cadre of Regular officers and Senior Ranks. While undoubtedly the UDR suffered over the years from a few 'rotten apples' with connections with Protestant paramilitaries, the vast majority of its members performed loyal and dedicated service, often at great risk to their lives, with over 150 killed, on or off duty, during 1970–1985.

Northern Ireland itself was soon organized into three brigade areas. While 39 Brigade looked after Belfast, 8 Brigade had responsibility for Londonderry and 3 Brigade the more rural southern half of the province. Tours in Ulster were, and still are, of three types. First, there are the so-called resident battalions, usually five in number, who live in the permanent barracks and do a two-year tour. There are also a number of battalions on roulement tours. Initially these were of four months' duration, but were extended to four and a half months in the early Eighties. Finally, there are shorter emergency tours, normally triggered by a sudden up-turn in violence. During the 1970s the roulement or Op BANNER tours came round with alarming frequency. To take The Green Howards as but one example, they served in Northern Ireland during the 1970s as follows:

June–September 1970	Belfast
July–November 1971	Belfast
October 1972–February 1973	Belfast
May–September 1974	Portadown
April–August 1975	Bessbrook
April–May 1976	Emergency tour
September 1978–March 1980	Aldergrove as a Resident Bn

It usually took a battalion two months' concentrated training to prepare for its tour, and, when leave before and after it were taken into account, it would mean that it would be diverted from its normal role for some eight months. This created particular problems in BAOR, from where troops of all cap badges had to be deployed, and meant that often a third of the major units within a division were either away in Northern Ireland, preparing for or recovering from it. Inevitably, this meant that mechanized warfare training suffered.

Many of the techniques and tactics which had been developed since 1945 in overseas counter-insurgency operations could not be applied in Ireland. Thus, when dealing with a riot there was no question of the 'shoot to kill' policy described earlier. Instead the rubber or plastic bullet and CS gas were employed, but in the early months, before these were introduced, the troops often just had to 'grin and bear it' as rocks and other missiles were thrown at them. As a Rifleman of the 2nd Royal Green Jackets recalled:

'I think the most intimidating thing about a crowd is that you can't get at them. You can take the abuse, I don't think they can overwhelm you as long as you've got the safety of the platoon around you and a back-up force behind, but if you get isolated, then you do get intimidated, and they can do it with bricks and bottles, or with the actual noise. As the crowd comes closer the chanting and howling gets deafening. . . .

'They were good; they could lob a fair-sized brick a hundred metres and hit what they were aiming at. They were bloody good, they'd had so much practice. I mean you'd stand there with your shield and they would get it onto the shield almost every time. We started with the four-foot shield but it wasn't very good at all, because if you lifted it up they took off your ankles, if you put it down they took your head off. So we went to the six-foots.'

During the 'shooting war', when IRA gunmen were nightly taking on Army patrols in the streets of Belfast, it soon became apparent that the standards of musketry had to be improved. Often the target was a fleeting one and in poor light, and so it was snap-shooting up to 200 yards which was especially important. The introduction of Electric Target, Crossing Target and Close Quarter Battle ranges helped to improve matters, as did the construction of 'tin cities', replicas of Ulster streets at Sennelager in Germany and Lydd in Kent, where troops were presented with realistic situations requiring immediate action. They were supervised by the Northern Ireland Training and Advisory Teams (NITAT), which consisted of officers and NCOs with extensive experience of the province and which maintained close liaison with Headquarters Northern Ireland in order to develop techniques for dealing with the ever-changing terrorist tactics.

It was soon found that the conventional section-sized 8–10 man patrol was too unwieldy for the urban jungles of Belfast and Londonderry, and instead the four-man 'brick' was adopted. This is easy to control and enables all members to be able to cover each other. If a brick came under fire, however, there was little it could do but seek cover and try to identify from where it was coming. Consequently, multiple-brick patrols were organized, with three or four bricks moving up parallel streets, ever ready to move swiftly to give close support to each other if trouble arose. Constant alertness was vital. The gunman's look-outs were always monitoring patrols, waiting for their attention to wander. Indeed, battalions were always at their most vulnerable towards the end of a tour, when they were thinking more and more about going home. They were also vulnerable at the beginning, when the IRA terrorist was keen to test their reactions. Every patrol was also, however, an information-gathering exercise as well as a deterrent, and this helped to maintain alertness.

In the rural areas the situation was different. While there was not the claustrophobia of the streets, the 'bandit country' of South Armagh and elsewhere was just as, if not more, dangerous. Indeed, because of the danger of mines and booby-traps, which necessitated lengthy route-clearance operations before vehicles could travel there, all resupply and major movement of personnel to and from the security force base at Crossmaglen became totally reliant on the helicopter.

In some ways Northern Ireland has been one of the sternest, and certainly the longest, test that the British soldier has faced since 1945. It became clear to the Army that there was no military solution to the problem and it was merely there to help contain the violence until a way could be found through the political morass. This could well have produced a serious demoralizing effect, but this was not the case. True, Senior Ranks, with up to five or six tours under their belts, tended eventually to get tired of it and frustrated by the lack of political progress, but young soldiers generally welcomed service in Northern Ireland and there is no doubt that it quickly matured them. As a sergeant in 1 Para said: 'You get nineteen-year-old lads going to Ireland and in those four months they gain ten years, ten years I'd say.'

A large part of this maturation was in learning self-restraint. The Army has been victim of a number of terrorist atrocities, from the cold-blooded murder of three young off-duty Royal Highland Fusiliers, two of them under 18 years old,* in March, 1971, through the Warrenpoint Massacre of August, 1979, when two massive bombs killed eighteen soldiers, mainly of 2 Para, but also the Commanding Officer of The Queen's Own Highlanders, Lieutenant-Colonel David Blair, to the bomb blast in August, 1988, which wrecked a bus carrying members of The Light Infantry back to their base in Omagh, killing eight and wounding twenty-eight. In the face of these the British soldier has shown remarkable restraint.

True, incidents like 'Bloody Sunday' in January, 1972, when 1 Para were accused of running amok, have caused people to question whether the Parachute Regiment has the right attitude for Northern Ireland because of its in-built 'go in hard' philosophy, but perhaps this should be balanced by incidents like that of a suitcase bomb dumped in the hallway of Springfield Police Station on 25 May, 1971. Realizing what it was, Sergeant Michael Willetts of 3 Para quickly ushered out two civilian adults and two children, shielding them with his body as they left and dying in the subsequent explosion, an act which earned him a posthumous George Cross. The vast majority of brutality accusations levelled at the Army are the result of terrorist propaganda, sometimes seized on by the media, as ever after a good story.

At base, though, the infantryman on the streets of Belfast was often put in an impossible situation. In his pocket he carried the now famous 'Yellow Card', which told him when he was permitted by law to open fire. Often, though, his decision had to be split-second, but if it was the wrong one he could find himself in court, accused of murder, as happened to Private Ian

---

* As a result of this, a ruling was made that no soldier under 18 would serve in Northern Ireland. This was not applied to other active theatres. The media made much of a 17-year-old Royal Scot as being the youngest British soldier to take part in the Gulf War in 1991.

Thain, a 19 year-old Royal Anglian, in 1984. If on a roulement or emergency tour, his living conditions were likely to be very cramped and his hours of work very long. Indeed, the only break the soldier got was five days' 'R & R' (Rest and Relaxation) leave midway through his tour. On the bonus side, apart from maturing the young soldier, Northern Ireland has proved to be ideal for honing basic infantry skills and for developing powers of leadership among young officers and junior NCOs. Indeed, it was very much a lance-corporal's 'war'.

In the early 1970s a new combat dress was introduced, with disruptive camouflage pattern. With it came a peaked combat hat, which some battalions adopted, although others preferred to retain the beret. For wear in barracks the latter half of the Sixties had seen more and more regiments adopt regimental sweaters, 'woolly pullies' they were nicknamed, and by the early Seventies the range of colours selected had become so wide that officialdom began to discourage them. Many infantry regiments therefore reverted in time to the Jersey Heavy Wool with reinforced shoulders and elbow pads. The old ammunition boot, with its leather sole and thirteen metal studs, which had served the British soldier for so long, gave way to the DMS (Direct Moulded Sole) version with rubber sole. A lightweight higher version, to give more support to the ankle, was also developed for urban patrolling in Northern Ireland. With the coming of the new boot the 1938 pattern web anklet also vanished, its place being taken by the short puttee.

In 1970, in an effect to boost recruiting, the Military Salary was adopted. The object of this was to make the serviceman more on a par with his civilian counterpart in terms of pay. Marriage allowance was removed and the single soldier's pay dramatically increased, although he now had to pay for his board and lodging. A particular element within the military salary is the so-called 'X' Factor, which reflects the unique disruption in a serviceman's or woman's life over that of the civilian, and which normally represents some 5 per cent of the salary. Accompanying this came a total switch to paying soldiers direct into their banks. Thus, gone was the traditional pay parade, with the soldier presenting his paybook to the paying officer, who then entered the sum to be paid, handed the money over, to which the soldier responded 'Pay and paybook correct, Sir!'

Montgomery's 1946 aim of improving the soldier's living conditions was also finally realized in the early 1970s with the launching of Op HUMANE. The barrack rooms were converted to single, two- and four-man rooms, complete with built-in cupboards, bedside lights, and even special boards for putting up pin-ups. The standard of army cooking improved to such an extent that the well-known cookery writer, Egon Ronay, declared in the early 1980s that it was generally significantly higher than that found in civilian catering.

In Germany more flexible forms of combat organization were being developed. In the mid-1960s battle groups and combat teams were created within each armoured and mechanized brigade. The former were based on the armoured regiment and infantry battalion headquarters. Each would normally command three squadron/company groups, the combat teams. These, in turn would be made up of a mixture of troops and mechanized platoons, which could be altered at will, depending on the tactical situation. The resultant closer bonding between infantry and armour was used by the proponents of the single combat arm as the main plank of their arguments.

Under the 1974 WIDE HORIZON restructuring exercise this flexibility in grouping of armour and infantry was taken a level higher. Brigades as such were abolished and in their place were introduced Task Force HQs in Germany, while the brigades elsewhere became known as Field Forces. The idea was that I(BR) Corps, which now consisted of three armoured divisions (1st, 2nd, 4th), each with a mixture of armoured regiments and mechanized infantry battalions, and having a number of Task Force HQs under command, could be tailored to fit any particular tactical situation.

The Task Force HQs also had a peacetime role as garrison HQs, but did not necessarily command the same units in the field as they did in barracks. This proved to be the main flaw in the concept in that the task force commander and his staff often did not know the units under his command as well as they might have done under the old system. Indeed, few outside the task forces themselves, which were designated by alphabetical letters, understood their composition or which divisions they served.* Consequently, at the beginning of 1981, the idea was scrapped and brigades were re-introduced.

During the Seventies the mechanized infantry battalions received enhanced weaponry. Their reconnaissance platoons, which had been mounted in Landrovers, were given the Combat Vehicle Reconnaissance (CVR) Scimitar variant with its 30mm Rarden gun. They also gave up their Wombat recoilless anti-tank guns for the Milan anti-tank guided weapon system, able to destroy any main battle tank up to a range of 2,000 metres.

The debate over the mechanized infantry's basic vehicle, whether it should continue to be a battlefield taxi or an armoured fighting vehicle as well, continued. In an effort to solve it, half the FV 432s were fitted with

* The author had personal experience of this. When, from his Ministry of Defence desk, he tried to find out which task force supported which division, he was told by the MOD branch responsible for army organization (ASD – Army Staff Duties) that they did not know.

Rarden 30mm turrets. This gave them the ability to knock out light armoured vehicles up to a range of 1,500 metres, and meant that some of the pressure, in view of the overwhelming Warsaw Pact numerical superiority in AFVs, could be taken off the tanks. Some infantrymen, however, were not happy about this, since to put armed APCs in the front line risked the danger that they would be destroyed. Thus, robbed of their taxis, the infantry would no longer be mechanized. The latter school won, and turrets were therefore removed, apart from on a small number of FV432s which were retained by the three infantry battalions who formed part of the Berlin Brigade.

The debate continued, however, and the upshot was that by the end of the decade a Mechanized Infantry Combat Vehicle (MICV), which could both act as a taxi and fight, was developed, based on the chassis of the Chieftain main battle tank.

Training in Germany, as elsewhere, followed the traditional cycle developed at the turn of the century of starting at the beginning of the year with individual skills and building up to formation exercises in the autumn. The latter meant that in Germany, at least, there was the opportunity to exercise over 'real' terrain, as opposed to being restricted to training areas, which were so churned up by tracks that they were either mud or dustbowls. By the end of the 1970s, however, the formation exercises were becoming increasingly restricted, both by the cost of damage caused and the growing 'greening' of Germany.

There were, however, opportunities for the troops to exercise elsewhere. During the 1960s this was Libya, but after Colonel Gadaffi assumed power the facility was transferred to Suffield, near Calgary in Western Canada, which troops from Germany still use. The great advantage of this is that, unlike in Germany, realistic 'live fire' exercises can be carried out.

Commitments overseas other than to NATO, or 'out-of-area', to use NATO parlance, continued to dog the infantry during the 1970s. There has always been one British infantry battalion stationed in Hong Kong alongside the Gurkhas. In early 1979 there was a sudden upsurge in the number of illegal immigrants (IIs) trying to get across the border from Red China into Hong Kong, the result of one of the periodic upheavals in the People's Republic. Soon all four battalions in Hong Kong were deployed on the border and an additional battalion had to be sent from Britain on an emergency four-month tour. A new continuous fence was constructed along the border, but it was not until October, 1980, when the Governor of Hong Kong put an end to the traditional 'touch base' policy, whereby an II could claim right of residence if he or she reached Kowloon and found accommodation, did the flood revert to its usual trickle.

The Turkish invasion of Cyprus in August, 1974, also caused a scare, and it seemed that at one point that the two Sovereign Base Areas were under threat. Certainly they had to cope with a flood of Greek-Cypriot refugees, but a negotiated ceasefire came into effect on 16 August and the tension was reduced.

Another area of concern was British Honduras, which became Belize in 1973 and attained full independence in 1981. Guatemala had long-standing territorial claims over the border region and throughout the 1960s a detached company from an England-based battalion was deployed in the colony. Guatemalan sabre-rattling in 1970 resulted in the Spearhead battalion, the Glosters, reinforced by elements of the Devon and Dorsets, being temporarily sent out as a deterrent. Thereafter the standing garrison was upgraded to a battalion reinforced with an additional rifle company. In 1977 tension increased once more when the Guatemalans actually moved troops across the border. The result was a further strengthening of the garrison to two complete battalions, supported by an armoured reconnaissance troop, 105mm light gun battery, air defence weapons, six Harriers, and a frigate. This defused the situation and after a time the forces were reduced to a single battalion, which continued to serve in Belize on a six-month tour, supported by the local defence force, which became increasingly effective over the years. One advantage that Belize did provide was the opportunity for infantrymen to operate in the jungle, something which had virtually ceased, apart from the Gurkha battalion stationed in Brunei, since the closing down of the Jungle Warfare School in Malaysia in 1971.

The British Army has always lived with the unexpected, but scarcely anything could have taken it and, indeed, the British Government so much by surprise as the Argentinian invasion of the Falklands at the beginning of April, 1982. The two formations that were immediately available for out-of-area operations were 3 Commando Brigade and 5 Infantry Brigade, the latter the successor of 16 Parachute Brigade. Since an amphibious operation was the only practical means, given the distances involved, of retaking the Islands, the force was built round 3 Commando Brigade, with 2 and 3 Para from 5 Infantry Brigade placed under its command. While the Falklands Task Force was being hastily assembled, the Ministry of Defence realized that additional troops might have to be deployed, if nothing else to act as a garrison on the Islands after they had been liberated. It was logical that these reinforcements should be built around HQ 5 Brigade, which was trained for out-of-area operations, but this only had one integral battalion left to it, 1st/7th Gurkhas. It might have made sense to call on The Queen's Own Highlanders, then the Spearhead battalion, which is kept on stand-by for rapid operational deployments, but they were primarily orientated towards Northern Ireland, and to replace them at short notice, which involved a handing over of stores and equipment, might have proved unnecessarily complicated. The other field formations in the United

Kingdom were dedicated as reinforcements for NATO, and to deploy them might have resulted in embarrassing questions being asked by Britain's allies. All that were left were the Foot Guards battalions on Public Duties in London, and it was to this source that the MOD turned.

On 5 April 2nd Scots Guards, at Chelsea Barracks in London, and 1st Welsh Guards at Pirbright were warned that their services might be required. At the time both battalions were deeply involved in preparing for the 'high season' of Public Duties, which begins with the Major General Commanding The Household Division's Inspection in April and culminates in the Trooping of the Colour in June. Four days later the Commanding Officers of the two Guards and 1st/7th Gurkhas were given an initial briefing by Commander 5 Brigade at Aldershot. It was, however, only on 14 April that the Guards battalions were told that, with effect from 19 April, they would be under command of 5 Brigade, and that on the following day they would be moving to the Brecon Beacons in Wales for two weeks' work-up training.

The Brecon Beacons had been chosen because of its rough similarity with the Falklands landscape. Here the battalions concentrated on physical fitness, field firing, tactical trainging and First Aid. Training in the last-named had undergone revolutionary changes during the past few years. Traditionally it had been the province of the company and battalion medical personnel, with the Drums having a wartime role as stretcher-bearers. Now the emphasis was on giving every man training so that he could apply immediate first aid to his wounded comrades.

In the meantime the Task Force sailed south and closed on the Falklands, having stopped briefly at Ascension Island. On 25 April came the first success, when M. Company 42 RM Commando recaptured South Georgia. This, though, was soon dampened by the loss of the first Harrier, over Goose Green, soon to become immortalized, and of HMS *Sheffield*, victim of an Exocet missile. On 12 May 5 Brigade set sail, the bulk of its men on board the luxury liner *QE2*, which, as one young officer commented, was 'a ridiculous way to go to war'.*

Nine days later 3 Commando Brigade made an unopposed (at least on the ground) landing at San Carlos and then began the battle to secure the beachhead in the face of persistent Argentinian air attack, which caused further casualties to ships. Pressured by Whitehall, Brigadier (later Major General) Julian Thompson, Commander 3 Brigade, gave orders to move out of the beachhead and begin the advance towards Port Stanley. His original intention had been that this should be done largely by helicopter,

---

* Among the many other ships taken up from trade was also the cruise liner *Canberra*, which carried the bulk of 3 Commando Brigade and then performed sterling service as a hospital ship.

but the loss of *Atlantic Conveyor* with three Chinook helicopters on board on 25 May put paid to this and the break-out had to be done on foot.

On 27 May 45 Commando and 3 Para began their advance and a new word was soon to enter the English language, 'to yomp', which means to march across country with a heavy load on one's back. The previous evening 2 Para began its march to the settlement of Goose Green and on 28 May fought its epic 15-hour battle there. This was the only action during the campaign which was fought largely by day and the lack of cover was one of the main reasons behind 2 Para's relatively high casualty bill.

The supreme self-sacrifice of Lieutenant-Colonel 'H' Jones after he had gone forward to get the attack going again after it had become stalled was typical of a commanding officer who had always led his battalion from the front and was later recognized by the award of a posthumous Victoria Cross to him. That his men, in spite of this blow, continued to fight their way forward and ultimately defeat the numerically superior forces opposing them is also a tribute to them and the Second-in-Command, Major Chris Keeble.

On 2 June 5 Brigade landed, their initial role being the security of the beachhead. Next day it was discovered that Fitzroy and Bluff Cove, 20 miles east of Goose Green, were clear of Argentinians, and Brigadier Tony Wilson, Commander 5 Brigade, decided to take immediate advantage of this and save his men a long march on their feet. He ordered 2 Para, which had now come under his command, forward to Bluff Cove, using the one surviving Chinook. This was successfully achieved and 2nd Scots Guards were embarked in the assault ship *Intrepid*, which took them part of the way to Bluff Cove. They had, however, to transship to landing craft off Lively Island, four hours' steaming time from their destination, because of the threat of a land-based Exocet system known to be near Port Stanley. After a nerve-wracking period when they thought that they might be engaged by an enemy warship (it turned out to be British), and amid rain and heavy seas, they arrived, wet and bedraggled, at Bluff Cove at 0545 hours on the 6th to be welcomed by 2 Para with hot tea. The landing craft then took the Paratroopers, who had subsisted for the last two days on mutton supplied by the Kelpers after their rations had run out, back to Fitzroy.

That night the Welsh Guards set sail in *Fearless*, but worsening weather and lack of landing craft necessitated half being landed on Lively Island and the remainder returning to San Carlos. On the night of the 7th/8th Welsh Guards tried again, this time on the Landing Ships Logistic (LSL) *Sir Galahad* and *Sir Tristram*, which also took the remainder of 5 Brigade. There followed, on 8 June, the air attack on the LSLs anchored in Bluff Cove, with the resultant loss of *Sir Galahad* with forty-eight of those on board killed and many others suffering appalling injuries, this in spite of a desperate barrage of small-arms fire put up by the Scots Guards. Of the dead, thirty-two were Welsh Guardsmen, mainly members of the Mortar Platoon.

Tragic though this incident was, and the recriminations continue over who was responsible for allowing the LSLs to remain in Bluff Cover during the hours of daylight, it could not be allowed to impede the advance on Stanley. The final obstacles which lay in front of the objective were a quarter-circle of hills, reading from south to north, Mounts Longdon, Two Sisters and Harriet. These guarded the west-east running Mount Tumble-down and, just to its north, Mount William. Three miles behind these lay Stanley itself.

By now 3 Para and 45 Commando had completed their yomp (the Paras preferred to call it 'tabbing'), covering some 50 miles, as the crow flies, over largely boggy and broken ground, each man carrying up to 120lbs on his back. They were now tasked with attacking Mount Longdon and Two Sisters respectively, while 42 Commando went for Harriet. These attacks were mounted on the night 11/12 June and were all successful, although 3 Para had an especially tough battle with the Argentine 7th Infantry, who were well positioned among the rocks and crags. Corporal Ian Bailey, a section commander, describes a critical point in the attack:

'Ian [Sergeant Ian McKay] and I had a talk and decided the aim was to get across to the next cover, which was thirty to thirty-five metres away. There were some Argentinian positions there but we didn't know the exact location. He shouted out to the other corporals to give covering fire, three machine guns altogether, then we – Sergeant McKay, myself and three private soldiers to the left of us – set off. As we were moving across the open ground, two of the privates were killed by rifle or machine-gun fire almost at once; the other private got across and into cover. We grenaded the first position and went past it without stopping, just firing into it, and that's when I got shot from one of the other positions which was about ten feet away. I think it was a rifle. I got hit in the hips and went down. Sergeant McKay was still going on to the next position but there was no one else with him. The last I saw of him, he was just going on, running towards the remaining positions in that group.'

McKay demolished a machine-gun nest, which was causing most of the trouble, with grenades, but was killed by a sniper in the process. His selfless action enabled the remainder of his company to get up on to the main ridge of the feature from where they could bring fire onto the remaining

Argentinian positions and resulted in the award of a posthumous Victoria Cross, the second of two won during the campaign. In all, 3 Para had twenty-two men killed and some thirty-five wounded.

Two nights later it was the turn of the Scots Guards. Their task was to wrest Tumbledown off the 5th Marine Regiment. Tumbledown is a narrow ridge-like feature approximately one mile long and covered with rocks and boulders. Lieutenant-Colonel (now Major General) Mike Scott planned to attack from the west and leapfrog his companies through the length of the feature. He also formed an *ad hoc* force from the Recce Platoon, Battalion HQ and A Echelon to mount a diversion to make the Argentines believe that the attack was coming from the south. This group of thirty men attacked ahead of the main body and became embroiled in a bitter fight against some sangars. They overran three, but then casualties began to mount and they were forced to begin to withdraw after two hours, only to find themselves having to negotiate their way through an enemy minefield. Their total losses were two killed and ten wounded.

The main attack was opened by G Company. The plan was for a silent assault until such time as the enemy opened fire. Thus, the Guardsmen wore berets to prevent their helmets from clinking during the advance, and, as G Company, who were to carry out the first phase, moved forward, there was merely desultory artillery fire, which included the odd starshell. G Company secured their objective without a shot being fired and Left Flank Company passed through them, bayonets fixed in the firm belief that they would not be so lucky. They were soon proved right. Heavy fire was opened and they were forced to go to ground. Gradually they managed to wrinkle out some of the Argentines from the rocks, using 66mm rocket and M79 grenade launchers.

Casualties, though, were beginning to mount. Worse, all but one of the battalion's mortars was out of action, the bipods broken and baseplates cracked because of the slope on which they were sited and the intensity of their fire. It also proved difficult to bring down accurate artillery fire because of problems in identifying which guns were firing short. The attack therefore ground to a halt. Eventually the supporting fire was sorted out and began to land on target. Left Flank skirmished their way forward, small groups covering each other with fire. Momentum increased, but at times the Guardsmen found themselves engaged in hand-to-hand fighting. The Company Commander, Major (now Brigadier) John Kiszely, having killed two men, and with no time to change his magazine, was forced to use his bayonet on a third. Guardsmen fell, dead or wounded, or became engaged in almost private battles against intransigent snipers and strong-points. Eventually Kiszely found himself on his objective, with just six of his men, three of whom were almost immediately badly wounded by machine-gun fire. As he himself said: 'A counter-attack, even if it had been two or three men in an *ad hoc* group with a corporal leading them, would have succeeded then.'

No such move on the part of the enemy was immediately forthcoming and gradually other elements of the company joined the small group on the objective, cheered by the realization that they could make out the lights of Port Stanley. A small counter-attack did now take place, but was driven off, but not without further casualties, including from a mortar bomb which killed two stretcher-bearers tending to the wounded.

Now it was Right Flank Company's turn. They had had a long cold wait while Left Flank fought its battle, and their Company Commander, Major Simon Price, was shocked to discover from Kiszely that they were fighting Regulars and not conscripts: 'I had gathered from the Commandos that all one had to do was to fire a few 66s and 84s and the Argentinians would come out with their hands up. This was clearly not so on Tumbledown.'

Price decided on a right flanking attack, but was told that he could not have any supporting artillery fire because it was believed that the Gurkhas,* who were to assault Mount William once Tumbledown had been secured, had got too far forward, which was actually not so. Consequently Right Flank had only its own weapons. They launched their attack and Lieutenant Robert Lawrence remembered shouting to his men: 'Targets fall when hit!', an allusion to firing against figure targets on the rifle range. The Argentinian fire was heavy, but the initial objectives were quickly taken, Lawrence, marvelling that his men followed him up on to a prominent ledge, leading his platoon to overcome a machine-gun post, killing two and capturing four. Pushing on again, another Argentinian machine gun gave trouble. Guardsman Pengelly:

'Tracer was streaming over our heads. I was a machine-gunner, but had already used up all my ammunition and felt a bit spare. I grabbed a grenade from someone and shinned up the Pinnacle. As I got to the top I tried to remove the pin, but it was bent and my hands were pretty numb from the cold.

Eventually I got it out and dropped it over the top. Stupidly I took a peek to see what it had done. Next moment I felt a blow in the thigh and toppled all the way down again. I'd still like to know who gave me the grenade with the bent pin.'

---

* The Commanding Officers of the Scots Guards and 1st/7th Gurkhas agreed to use the cries of 'Hey Jimmy!' (a well known Scottish cry) and 'Hey Johnny!' (from the British soldier's traditional nickname for the Gurkhas, 'Johnny Gurkha') to identify each other in the dark. These had the added advantage that the phonetic 'J' is unpronounceable in Spanish.

Pushing forward, using skirmishing or 'pepperpotting', as it was popularly known, tactics, Right Flank managed to clear the remainder of Tumbledown and it was secured as dawn appeared. This was not before Lawrence was hit in the head, suffering a wound that few believed he would survive; he did so, even though he lost part of his brain. The Scots Guards' efforts won a DSO for their Commanding Officer, MCs for Kiszely and Lawrence, two Distinguished Conduct Medals, two Military Medals (one to Guardsman Andrew Pengelly), and fourteen Mentions in Despatches. The Gurkhas moved forward to seize Mount William, but the defenders fled, victims of their own propaganda on the Gurkhas' ferocity. This denied the Welsh Guards, now reinforced by two companies of 40 Commando, their opportunity to revenge Bluff Cove with an attack on Sapper Hill just in front of Stanley.

The same night as the attack on Tumbledown 2 Para had been called to attack once more, the only battalion to make two major attacks during the campaign. This time their objective was Wireless Ridge, which lay to the north of Stanley. They had, in the meantime, received a new commanding officer, Lieutenant-Colonel (now Brigadier) David Chaundler, who had been removed from his MOD desk, parachuted into the sea 100 miles north-east of the Falklands, picked up by the frigate *Penelope*, and joined his battalion four days after the death of 'H' Jones. They got quickly up on to the ridge, but clearing it proved more difficult. Not for the first time in the campaign were Milan anti-tank guided missiles used to destroy strongpoints. Nevertheless, 2 Para succeeded at the cost of three men killed. Chaundler, comparing the action with Goose Green, commented:

'Because we had Goose Green behind us, there was a different attitude in the battalion and, while the soldiers were more apprehensive, they were also much more professional because they knew what was going to happen.'

The successes of this night convinced the Argentinians that there was now nothing to stop the British from achieving their objective and they surrendered on the night of the 14th/15th.

In the context of the 'hi-tech' style of war which the British Army was now training for in the NATO context, the Falklands campaign was in many ways a throwback to a bygone era when the basic element was the infantryman on his feet with his rifle and bayonet. The night battles in the hills were very much reliant on the section commander and the individual rifleman, and it is very noticeable that it was the initiative shown by Junior Ranks which so often turned the tide. This reflected the high quality of their training and the experience gained in Northern Ireland. Indeed, such was their performance that they surprised their seniors. One senior NCO of the Scots Guards described the young Guardsmen as being 'like tigers' on Tumbledown and another said: 'We were very impressed with the youngsters'.

There was also a stark contrast on the two sides in the relationships between officers and men. The Argentinian forces were mainly made up of conscripts, whose officers kept themselves remote from them, even to the extent of having different rations. In contrast the members of the British battalions and commandos had grown up together and were, of course, all volunteers. Thus, in 2nd Scots Guards, for example, two of the company commanders had been platoon commanders in Colonel Mike Scott's company and some of the warrant officers had been his platoon sergeants. This had bred mutual respect and friendship. Indeed, the Battalion Medical Officer, Lieutenant-Colonel Warsap, commenting on the marked absence of battle shock as a result of Tumbledown, firmly believed that the reason for this was the highly disciplined regimental atmosphere, combined with a distinct ethnic character and the feeling of the Battalion as a family. He also stressed the overriding respect for the individual which existed, something that perhaps might surprise those who tend to think of the Foot Guards as mere mechanical parade ground toys. Thus, the Falklands reinforced the arguments of those fighting to preserve the traditional regimental system, something which within the next decade would once more come under severe threat.

CHAPTER FOURTEEN

Present and future
1982–1994

THE euphoria with which the British Public greeted the Falklands Task Force on its return home surprised those who had taken part in the campaign, and much of the world at large. Indeed, it had been many years since the stock of the British fighting man had stood so high. A plethora of awards and decorations was bestowed by a grateful nation, and, for the first time since Korea, battle honours were awarded. The Scots and Welsh Guards, Parachute Regiment and 7th Gurkha Rifles received 'Falkland Islands 1982', and honours for specific actions were granted to the Scots Guards ('Tumbledown Mountain') and Parachute Regiment ('Goose Green', 'Mount Longdon', 'Wireless Ridge'). But while the Foot Guards returned to Public Duties and the Paras resumed their normal duties in Aldershot, the Falklands needed to be very much more strongly garrisoned than it had been prior to the Argentinian invasion. Consequently an infantry battalion group was based there, together with air and naval elements. Battalions did a six-months' tour, which included having a detachment on South Georgia. In recent years, however, the garrison has been reduced to company strength.

The Falklands campaign was generally considered to be a 'one off', something which was most unlikely to occur again, especially in view of Britain's prime commitment to NATO. Thus, the Army's main attention continued to be on I(BR) Corps in Germany. In 1980 and 1984 two very large-scale exercises, CRUSADER and LIONHEART, were conducted by the Corps. What marked these as different from the normal manoeuvres in Germany was that they included major reinforcements from Britain taking part. Not only was 2nd Infantry Division, responsible for the security of the Corps rear area, brought across complete, but so were other formations from Britain whose role was to reinforce the three armoured divisions in place (1st, 3rd, 4th). For LIONHEART the reinforcement phase was taken one stage further, with 4,500 reservists also being called up to take part.

This exercise also revealed a new range of equipment for the infantry. First and foremost, on trial with the Irish Guards was the long-awaited

MCV-80, the Mechanized Infantry Combat Vehicle, which was to replace FV432, at least in part. New webbing was also on show, as well as LAW-80, a new hand-held anti-tank weapon, latest in the line that had begun with the PIAT.

Soon to come into service, too, was a new family of small arms, SA-80, with a calibre of 5.56mm. This has two members. The first is the Individual Personal Weapon, which has replaced the SLR, and in doing so has brought back the slope arms into the drill book, although SA–80 is uncomfortable to hold in this position for any length of time and consequently a new drill movement has had to be introduced for changing shoulders at the slope. The other member is the Light Support Weapon, which fulfills the function of the light machine gun as the infantry section's intimate fire support. The GPMG is, however, retained in the sustained fire role and continues to use 7.62mm ammunition, because of the longer effective ranges required by this type of weapon. What is significant about SA-80 is that it is very similar in appearance to the RSAF Enfield concept for a 0.280-inch weapon at the end of the Forties.

Another significant change was the introduction of a new boot. The DMS version had proved increasingly unpopular, mainly because it was not waterproof. Indeed, many troops in the Falklands wore civilian hiking boots in preference, and among those who wore the DMS boot there were some forty cases of trench foot. A new boot was, however, under development by the Army's Stores and Clothing Research and Development Establishment (SCRDE). This is higher, reaching above the ankle and has the tongue sealed to the inside of the boot in order to prevent water seeping in through the lace holes. An added advantage is that puttees no longer have to be worn with it, trouser bottoms merely being tucked into the top.

A new helmet also came into service in the mid-1980s. Its main advantage is that, made of kevlar, a very tough plastic, it is much lighter and more comfortable than the traditional steel variety, but gives just as good, if not better, protection. It is now standard that the helmet is worn at all times on operations and training in the field. This makes sense in terms of individual protection, but there is a disadvantage. The sight of helmeted soldiers with blackened faces in Northern Ireland did serve to depersonalize them, which acted against any residual effort towards winning 'hearts and minds' in the province.

Exercise LIONHEART also put on public view a new type of infantry. The concept of airmobile infantry, moving about the battlefield in helicopters, is a descendant of the airlanding troops of the Second World War. The French first developed the idea in Algeria in the late 1950s, and the Americans made extensive use of airmobile troops in Vietnam. The Warsaw Pact also introduced them, as did the Germans. The British were therefore somewhat late on the scene when it was decided in 1983 to set up a trial. For this purpose one of the armoured brigades in Germany, the 6th, became 6 Airmobile Brigade. It consisted of two battalions, 1st Gordons and 1st Light Infantry, supported by a field regiment RA, field squadron RE, and Army Air Corps squadron of Gazelle scout and Lynx anti-tank helicopters. The concept was that this force could be speedily deployed to block a Warsaw Pact thrust which looked like breaking through the main defensive position. To this end each battalion was given no less than forty-two Milan ATGW systems, organized in three platoons, one for each rifle company. The Chinook and Puma helicopters which were to provide the lift for the battalions were supplied by the RAF.

The airmobile battalions could also be used in the advance, as an officer of the 1st Light Infantry relates in a description of the part played by his battalion on LIONHEART:

'At the weekend the Battalion went on the offensive. The recce platoon was clandestinely inserted by Lynx some 30km behind the enemy FEBA [Forward Edge of the Battle Area]. It pinpointed enemy positions and located helicopter landing sites. 24 hours later a Brigade operation was mounted to seize bridges ahead of the ground advance. C Company LI led a massive armada of helicopters carrying the bulk of the Brigade before last light. After dark the remainder of the Battalion found itself in two villages, surrounded by marauding enemy armour. In daylight, fierce anti-armour battles splintered into exhilarating street fighting while ammunition was brought in by Puma and casualties evacuated on returning aircraft. A subsequent assault by battalion strength enemy saw 28 AFVs killed by the Battalion and our affiliated guns engaging Leopards over open sights.

'As a final act, the Battalion again flew forward to secure routes through close country for advancing armour. A rifle company inserted at last light created havoc among the enemy hides and resting tanks. At dawn the remainder of the battalion flew forward in one wave of aircraft to reinforce them. The tactical flying of the Chinook pilots resulted in the bulk of the Battalion being successfully deposited into the midst of an armoured mêlée.

The trial was extremely successful and resulted in the setting up, in 1988, of a permanent airmobile formation, 24 Airmobile Brigade, which is based in Yorkshire.

Mention of fighting in built-up areas leads on to another aspect of infantry tactics which was resurrected from the past during the 1980s. The growing conurbanization of the Federal Republic of Germany resulted in increasing realization among the planners that, in the event of a Warsaw Pact attack across the Inner German Border, the likelihood that British troops would find themselves involved in fighting in towns and cities was growing, Consequently the street fighting techniques of 1939–45 were dusted off and FIBUA (Fighting in Built-Up Areas) became an important part of the Infantry's training. Replicas of German villages were constructed at Sennelager in Germany and on the southern edge of Salisbury plain, near Heytesbury. Unfortunately, those responsible for siting the latter seem to have paid little attention to the local environment and the result is that it sits up like a sore thumb amid the surrounding countryside. Nevertheless, today's infantrymen are now relearning the lessons of 1944–45, especially how arduous and lengthy a business urban fighting is and how high is the rate of ammunition expenditure.

A more abstract concept developed by the British Army in the 1980s was that of the 'One Army'. The message behind this was that the Regular and Territorial Armies were as one, recognizing that in Germany, at least, I(BR) Corps would find it difficult to function in war without its TA reinforcements. The growing importance of the TA had been recognized in 1981, when the government announced that its strength would be increased from a target strength of 73,000 to 86,000 by 1990. For the TA infantry this meant additional battalions, increasing the number to forty-one. The beneficiaries of this were not necessarily the existing TA regiments, an admission that the creation of regiments with brigade titles had been a mistake. Instead, the following Regular regiments were given a TA battalion:

The Staffordshire Regiment – 3rd Bn
The Cheshire Regiment – 3rd Bn
The Devon and Dorsets – 4th Bn

The Yorkshire Volunteers were to raise a 4th Battalion, as were the Light Infantry, and the Royal Green Jackets a second TA battalion. Finally, one entirely new TA regiment was raised in London, the 8th The Queen's Fusiliers. Some cynical commentators viewed the selection as to where the new battalions were to be based as being heavily influenced by marginal Conservative seats, while others saw it as a plot to develop defence on the cheap by relying on part-time rather than Regular soldiers. Even so, this expansion provided the TA with a significant boost to morale.

One curious footnote to the 'One Army' concept was that, at the same time as it was introduced, the listings of TA units and their officers in the annual *Army List* were banished to the back of the book instead of being shown alongside their Regular counterparts as had been the case since the formation of the Territorial Force early in the century. This appeared to be in total opposition to what 'One Army' was supposed to stand for.

The dramatic events of 1989 in Eastern Europe, culminating in the opening up of the Berlin Wall on 9 November, which brought an end to the Cold War that had dominated Europe for over forty years, left NATO in a vacuum. The cornerstone of many of its member nations' defence policies had suddenly been removed and, while the collapse of the Communist monolith was warmly welcomed, the crystal ball had clouded over. One thing was certain, though. There was no longer the requirement to have such large ground forces stationed in the NATO Central Region. Hence governments could reduce their defence expenditure. In the 1990 Defence Estimates the Secretary of State, Tom King, announced that there would be an extensive defence review entitled Options for Change. This sent a shiver through the regiments since it was seen as inevitable that the number would be reduced. Worse, the regimental system itself was once more being brought into question.

New factors had come into play which strengthened the hand of those, both inside the Army and out, who believed that the system was outdated and hopelessly inefficient. A falling-off in recruiting during the mid-Eighties, a reflection of lower unemployment figures, had left some Regular battalions well below strength, a few even being forced temporarily to disband rifle companies.

There was also growing concern about the quality of recruits coming into the army. Standards of physical fitness had fallen drastically, and, having worn nothing but heelless trainers all their lives, new recruits found it difficult to adjust to wearing the Army boot. Their view of the world seemed to be entirely conditioned by Ramboesque videos, and, more seriously, they had an inbred selfishness and disregard for traditional values which appeared to be in opposition to the established concept of The Regiment. One apparent symptom of the frustration felt by instructors was a flood of media stories of bullying, especially at infantry depots, and there was also an alarming rise in the number of recruits who were opting to leave during their basic training. So serious did this become that in Autumn, 1989, new guidelines for recruit training were issued. From now on the traditional approach of stripping away the recruit's civilian character and rebuilding it was to be tempered by a much more gentle approach. Military discipline was now to be introduced gradually, with instructors being told that they must not shout at their recruits during the first three weeks of training, and to counsel them rather than place them on a charge for misdemeanours.

In the midst of the discussions on what shape the post-Cold War British Army should be there was a crisis in the Middle East. On 2 August, 1990,

Saddam Hussein's forces invaded Kuwait and annexed it. Within a week US forces began to deploy to Saudi Arabia, to be followed shortly by contingents from the Arab League. The British Government, too, quickly deployed naval and air elements to the Gulf and prepared to send ground troops as well. The initial reaction was to look to the two formations that were geared to out-of-area operations, 3 Commando and 5 Airborne Brigades, but it quickly became clear that troops operating on light scales were not what was required against an enemy known to have a large number of tanks and other armoured vehicles in a desert arena. Consequently it was decided in September to send an armoured brigade from Germany. Thus the mechanized battle for which BAOR had trained for so long looked as though it was finally about to come to pass, but on terrain very different from the North German Plain.

The brigade selected, 7 Armoured, still had the Jerboa that its illustrious predecessor of the Desert, Burma and Italian campaigns of 1939–45 bore as its formation sign. It consisted of two armoured regiments, Royal Scots Dragoon Guards and Queen's Royal Irish Hussars, equipped with the Challenger tank, and a mechanized infantry battalion, the 1st Staffords, with Warrior, as MCV–80 was now called. In order to bring all three up to war establishment they had to be reinforced from other regiments and battalions. The Staffords received a complete company of the 1st Grenadier Guards. This appeared to make a complete mockery of the regimental system, highlighting the inability of regiments to bring themselves quickly up to strength from their own resources. This was not so, according to the Brigade Commander, Brigadier Patrick Cordingley:

'We used to think of this system making regiments or battalions fight more vigorously because of the brotherhood links, but in this case it worked almost in reverse. For example, when a Grenadier Guardsman came to join the Staffords he was treated as an honoured guest and looked after exceptionally well. In return, he made certain that he was not going to let down his parent battalion.'

After some work-up training in Germany 7 Armoured Brigade flew to Saudi Arabia and began to unload its armoured vehicles, which had been moved by sea, on 16 October. The Brigade wsas initially placed under command of 1st US Marine Division. In November the decision was taken to increase the size of the British contingent in Saudi Arabia to a complete division. To this end HQ 1st Armoured Division was deployed, together with 4 Armoured Brigade, another formation which had the desert rat as its symbol in recognition of the brigade of the same title which had formed part of 7th Armoured Division during the Second World War. This consisted of two mechanized battalions – 1st Royal Scots and 3rd Royal Regiment of Fusiliers, both also equipped with Warrior – and the 14th/20th Hussars in Challengers. These battalions, too, had to be reinforced by a further Grenadier company and elements of The Queen's Own Highlanders. Shortly before the ground war erupted three further battalions – 1st Coldstream, 1st Royal Highland Fusiliers, 1st King's Own Scottish Borderers – were sent out to the Gulf primarily for POW-handling duties, but also to provide casualty replacements, as were 1st Scots Guards.

By the end of November, 1990, the initial Coalition's strictly defensive posture had been transformed as plans were developed for the liberation of Kuwait. In the meantime training was the main priority. While many elements in 1st Armoured Division had recently visited the Suffield training area in Canada, whose stark open terrain was not too dissimilar to the desert, there still remained much to be done to prepare for combat. Vital for the instilling of self-confidence was to practice attacks on trench and bunker systems using live ammunition, but without the usual peacetime safety restrictions. In other words it was to employ the same techniques used at the battle schools of the Second World War. The troops, understandably, initially approached these exercises with not a little trepidation. But after a while, as a troop leader in The Royal Scots Dragoon Guards noted:

'The attacks gave us so much confidence in the Staffords. Watching them with live ammunition, what they were doing was unbelievable. The amount of ammunition fired was unbelievable. The assault itself was nothing short of spectacular. One of the troop sergeants said that as they actually went in with bullets flying all over the place, he stuck his head out of his hatch to try and get a clearer idea of where he was. He said that that was the last time he would do that, the whole thing was quite terrifying. . . . If we had to go in somewhere for real, we knew that the Staffords would just rip the enemy apart.'

The Commanding Officer of The Royal Scots, Lieutenant-Colonel Iain Johnstone, also commented: 'We needed to show everyone we meant business and the Jocks thrived on it. They took a pride in the tough image they were creating and as the Press began to notice, the tougher they became. They positively revelled in it.' This confidence continued to increase and helped to keep morale at a high level. Indeed, given the host nation's insistence that the troops strictly observe the country's religious laws, which meant no alcohol and no pin-ups, and even no religious Christmas cards, and that there were precious few recreational amenities to be had, morale could have been a problem. Hard training and close-knit comradeship meant that it was not. As for the ban on alcohol, many commanders commented that their soldiers benefited from it, becoming noticeably fitter, and the incidence of petty military crime showed a marked decrease.

When DESERT SHIELD finally became DESERT STORM in the early hours of 17 January, 1991, with the first air strikes being mounted against targets in Iraq and Kuwait, there was relief that the waiting was likely to be over soon. The immediate gaining of air supremacy and the reports of the growing damage being done to Iraq's military infrastructure served to maintain the confidence of the troops on the ground, but as the air campaign continued there was plenty to do. Most important was the switch of command of 1st Armoured Division from the US Marines to US VII Corps, the main armoured element of the US forces in the Gulf. This involved moving the complete division, including its logistics, a distance of over 100 miles, an operation which went surprisingly smoothly. In the two weeks before Operation DESERT SABRE was launched the Divisional artillery had the opportunity to flex its muscles, carrying out 'shoot and scoot' engagements while the armour and infantry made their final preparations.

The ground force offensive itself was launched before dawn on Sunday 24 February. The greatest immediate threat was that the Iraqis would employ chemical weapons and all troops advanced wearing their NBC (Nuclear, Biological and Chemical) suits, with flak jackets worn over the top. Medical facilities had also been prepared for a large ingress of chemical casualties, the 1st Royal Scots battle group alone having three doctors and nineteen ambulances. Luckily the Iraqis never used their chemical arsenal. The Division passed through the breach created by the Americans in the border defences some twelve hours ahead of schedule, and then began a four-day high-speed Blitzkrieg, which was to take it across almost the whole breadth of Kuwait, a distance of some 200 miles.

For the infantry in their Warriors the fighting, especially by night, was one of stark contrasts. Iain Johnstone of The Royal Scots: 'Fighting at night in particular, using the II [Image Intensification for night vision] sights, was like playing a computer game on a green monitor with the sound turned down. It was easy to be calm because there was an unreality about it all. It was only when the hatch was thrown back to debuss that there was a shock wave of sound and tracer bouncing off everywhere. You felt safe – too safe – and detached from reality. Not an Infantry feeling at all.' As for the attacks against Iraqi positions, here is a description of a typical one, carried out by a squadron of The Royal Scots Dragoon Guards and a company of the Staffords:

'It was a particularly unpleasant night; it was raining quite heavily and visibility was down to about 15 metres before you could see anything the size of a Warrior. It was absolutely black. Thirty seconds before we went in the tanks opened up, and when the vehicles they hit started burning the infantry had a reference point to aim for.

'The attack was brilliant. All the drills worked. On the objective there was the main command site and then, off to the right, an area of bunkers and trenches. These objectives were allocated over the radio and the tanks and platoons split to them about 300 metres out.

'And when the infantry debussed and stepped into the blackness it was a step into the unknown for them. They had to overcome fear as they wrenched themselves away from the Warriors and went down and crawled forward desperately looking for a target and then a sprint forward with rain and sweat in their eyes and down again before they got to the bunker areas. Bullets, both friendly and enemy, seemed to be flying everywhere. Private Evans' life was saved when an AK47 bullet lodged in a rifle magazine in his breast pocket. We had then constant debussing and embussing. 1st Platoon re-embussed and went to the right where they took out a command bunker. We then re-embussed 3rd Platoon after a brief fight and put them and two tanks through the centre to another position and there they sustained three casualties. We also had another tank and one of the Milans grouped together putting down fire support as that platoon ran in. As soon as another position was identified, fire was put down. You could see where the tracer was going and where it was being fired from. When we were sure we weren't firing at each other we would move on again. Some of the assaults were very tight and it was undoubtedly a concerning time. The potential for running one tank/platoon group into another was enormous.

'The squadron and the company regrouped swiftly and although we had taken five casualties we all knew that whatever else happened we had done it and despite atrocious conditions it had worked.

The secret of success in these conditions was well practised battle drills, which enabled everyone to do the right thing instinctively.

There was though a penalty to be paid for these non-stop rapid operations. In spite of the adrenalin produced by a sense of mounting success, fatigue became increasingly apparent. Actions, both physical and mental, slowed, and small mistakes began to creep in. The vast destruction being wrought on the demoralized Iraqis also began to have a depressing effect on morale, especially during the Division's final action, the cutting of the Kuwait City to Basra road, which subsequently entailed clearing a route through the devastation caused by Allied airpower on the Iraqis fleeing from Kuwait City.

As for British casualties, these remained remarkably light, and it was tragedy that the highest loss should have been as a result of a US A–10 ground attack aircraft knocking out two of 3rd Royal Regiment of Fusiliers' Warriors, killing nine men and wounding thirteen others.

What compounded this still more were the subsequent attempts by lawyers representing the next-of-kin to find a scapegoat for it. The Oxford Coroner's Court verdict of unlawful killing reflected a total lack of understanding of what happens on the battlefield in that, in spite of the 'hi-tech' nature of modern warfare, the 'fog of war' still persists and the stress of combat can result in mistakes being made, regrettable as they are. Celebration of the verdict by one of the solicitors producing champagne on the steps of the court was at best in very poor taste and merely served to emphasize the ignorance of many as to what war is about.

In spite of this and other cases of casualties being inflicted by 'friendly fire', there is no doubt that the Division's performance had been spectacular, even given the poor state of its opponent. This is to the great credit of its members, especially the tank and Warrior crews, and did much to lay to rest, as the Falklands had nine years earlier, the concern over the lower quality of recruits joining the Army.

Throughout the deployment of the Gulf and subsequent fighting the review of the Armed Forces had continued, although some had hoped the Gulf War might provide at least a partial stay of execution. With an easily identifiable potential enemy no longer in existence, it was not easy, either for Britain or NATO, to formulate a coherent defence strategy to fit the immediate post-Cold War era. Consequently restructuring was heavily influenced by economic considerations. For the Regular Army it meant a reduction in strength from 156,000 to 110,000. The Royal Armoured Corps, realizing that a large reduction of strength of the forces stationed in Germany, where the bulk of its regiments were based, was inevitable, was prepared for loss through amalgamation of a large number of them – eight out of nineteen as it turned out. For the Infantry, however, the losses were to seem just as high, with the effect on the divisions as follows:

	No of Battalions	
Division	1992	1995
Household	8	5
The Queen's	9	5
The King's	8	7
The Prince of Wales's	9	6
Scottish	7	5
Light Division	6	4

The Parachute Regiment was to retain its three battalions, while the Gurkhas would be reduced to a single large regiment of three battalions. Thus, by 1995, when restructuring was completed, the infantry would have been reduced from fifty-five to thirty-eight battalions, reflecting a strength which had not been so low since the reign of George I.

Unlike previous exercises of this type, the decision over which regiments were to amalgamate was not determined primarily on seniority within each division. The key factor was to be current strength, with the weakest going to the wall, tempered by spreading the burden throughout the six administrative divisions. For The Royal Scots, who, as the oldest and most senior regiment of the Line, had long felt their position in the Army List to be sacrosanct, it was a particular shock to be informed, shortly after their return from the Gulf, that they were to amalgamate with The King's Own Scottish Borderers. The Queen's Own Highlanders were to feel bitter for a different reason. The product of the 1961 amalgamation between the Camerons and Seaforths, they were now to amalgamate yet again, this time with the Gordons, while the other two Highland regiments, The Black Watch and Argylls, were preserved in their pure state. Indeed, it was in Scotland that the outcry against the cuts was at its most vociferous. It became an issue during the April, 1992, General Election, with the Scottish National Party vowing to preserve all the Scottish regiments intact.

But elsewhere there was also much unhappiness. It has been said that Her Majesty The Queen was much disturbed by the reductions to the Household Division, with the three second battalions of Foot Guards being placed in suspended animation. This meant that from 1993 there were only three battalions on Public Duties, which meant an inevitable reduction in the extent of these. With the announcement in October, 1992, that the three would receive an increment of an additional one hundred men each, the Public Duties commitments have not, however, been as much reduced as many feared. True, the Trooping of the Colour is now performed by six 'Guards' rather than eight, but this was the case in 1982, with 2nd Scots Guards and 1st Welsh Guards away in the Falklands. The guard at the Tower of London, seen as the most likely victim of the cuts, has been preserved, albeit at a lower strength, and the ancient Ceremony of the Keys continues to be carried out nightly.

In England the most startling amalgamation, one which was brought forward from early 1993 to July, 1992, was that between The Queen's Regiment and the Royal Hampshires. Not even the fact that it was a large regiment of three Regular battalions could save The Queen's from losing their identity; they suffered the penalty for having had a poor recruiting record over recent years. The Royal Hampshires were equally unhappy, not least over losing their pure territorial title, the resultant regiment, with just two Regular battalions, being called The Princess of Wales's Royal Regiment (Queen's and Royal Hampshires). In fact, it was southern England and the Midlands which suffered most, with the regiments of Yorkshire and Lancashire escaping unscathed.

In Ireland, largely for political reasons, it was decided to merge the Ulster Defence Regiment with The Royal Irish Rangers in order to bring it more into the mainstream of the Army. Instead, however, of inventing a new title, the Colonel of the Regiment, Major General Roger Wheeler, managed to obtain agreement for an old title to be resurrected. The one he selected was highly appropriate, The Royal Irish Regiment, held until 1922 by the most senior of the Irish regiments, the old 18th Foot of Namur fame. The new regiment also came into being in Summer, 1992, with one Regular battalion, the remainder being TA and UDR battalions. The upshot of all these amalgamations was to mean that from 1995 the senior unadulterated regiment of the Line would have been The Green Howards, with The Royal Welch Fusiliers, who managed to ward off a proposed amalgamation with the Cheshires, being the only other regiment still able to claim an untainted ancestry throughout its existence. Of the Cardwell creations only The Duke of Wellington's, Black Watch and Argylls would remain pure. Yet, in spite of fears in some quarters and pressures in others, the regimental system survived, at least for the time being.

Deeper concern continued to be expressed, however, over the danger that the cuts had gone too far and that the Infantry might find itself stretched as never before. True, its commitment to NATO in Germany was now being significantly reduced, with the Berlin garrison, which still accounted for two battalions in 1992, closing down when the last Russian troops left the eastern part of Germany in 1994, and with only one standing armoured division, the 1st, remaining in Germany, as part of the newly formed Allied Rapid Reaction Corps (ARRC).

The 3rd Division, however, based back at its traditional home on Salisbury Plain has a reinforcement commitment to NATO. Also, while the battalion based at Gibraltar was withdrawn in 1989 and that based in Hong Kong will have to withdraw by 1997, together with the Gurkhas, overseas commitments remained. Cyprus still accounted for three battalions (one UN, two in the Sovereign Base Areas) and the Falklands continues to require a garrison for the foreseeable future. Furthermore Northern Ireland still had some twelve battalions on average, half of them resident. This meant just under one third of the Infantry, at its post Options 1995 strength and including the Gurkhas who could not, for political reasons, be sent there, in the province at any one time. When pre-Northern Ireland training and post-tour leave was taken into account, this proportion was even larger.

The fear was that existing defence commitments would both leave very little infantry in reserve and stretch existing resources to breaking point. An indicator that this might already be becoming so was the Government's hesitation in committing ground forces to serve under the UN flag in strife-ridden former Yugoslavia. But the 1st Cheshires, reinforced by a company of the 1st Royal Irish and elements from other battalions, and supported by a 9th/12th Lancers armoured reconnaissance squadron, were deployed there in Autumn, 1992, not to keep the peace in the traditional UN sense, but to provide escorts for humanitarian aid convoys. At the same time the British Government had to turn down a request from the Americans to assist in their military operation to overcome the famine and anarchy in Somalia, a tacit admission that the post-Options Army was becoming dangerously overstretched.

The high media profile enjoyed by the Cheshires in former Yugoslavia, heightened by their first fatality, Lance-Corporal Wayne Edwards attached from the Royal Welch Fusiliers, who was killed at the controls of his Warrior in crossfire between Bosnian Muslims and Croats, merely served to increase the general belief that the Army was being pared too much. This culminated towards the end of January, 1993, in the widely leaked and highly critical conclusions of the House of Commons Select Committee on Defence on the Options for Change exercise.

As Northern Ireland had in 1971, Bosnia provided the critical straw for the camel's back, and on 3 February, 1993, Secretary for Defence Malcolm Rifkind announced in the House of Commons a partial U-turn. The Army would be allowed an additional 3,000 men on top of its post-Options strength of 116,000 personnel. Part of this would allow existing units to be brought up to strength for operational commitments without having to be reinforced from elsewhere, but, more important, two additional infantry battalions were to be retained in the order of battle. As a result the amalgamations between The Royal Scots and King's Own Scottish Borderers, who had already agreed on a new tartan, cap badge, and title of the Royal Scottish Border Regiment, and that between the Cheshires and the Staffords were cancelled. Celebrations among the four were instant, with Lieutenant-Colonel Bob Stewart of the Cheshires telling a BBC Radio interviewer that evening that the 'last drops' of champagne in Bosnia had been all but consumed and would be before the night was out.

As for those regiments – The Gordons, Queen's Own Highlanders, Glosters, Duke of Edinburgh's Royal Regiment – who still faced amalgamation, Rifkind's change of heart merely served to give campaigns against the mergers fresh impetus. This was especially so since overstretch had not been overcome. The Prince of Wales's Own Regiment of Yorkshire relieved the Cheshires in Bosnia in May, 1993, less than a year after they had returned to Germany from an operational tour in Northern Ireland, and the same was to apply to the Coldstream Guards, who went to Bosnia in November of the same year.

The Territorial Army underwent a similar exercise to its Regular counterpart. In some ways the problem was more complex because there has always been much local vested interest, especially political, in the TA, and the territorial links which battalions have are even more tightly bonded. At the same time, apart from the inevitable reduction in battalions with a NATO reinforcement role, the recognizable importance of home defence in the post-Cold War era was much diminished. The solution eventually adopted was to give priority to reducing companies rather than battalions, giving the latter a new and smaller three-rifle-company organization and reducing the number of support weapons. Thus, while the number of rifle companies was reduced from 164 to 109, only five battalions were lost. Although in reality six were axed, a new one-battalion TA regiment was formed. This brought back a famous name into the TA order of battle, The London Regiment. Its headquarters was created from that of the 8th Queen's Fusiliers, which has been disbanded, and its three rifle companies were formed from those which respectively belonged to the Fusiliers, The London Scottish and The London Irish, the latter two companies having been outstations of the 1st/51st Highland Volunteers and The Royal Irish Rangers (V). The three companies do, however, retain their distinctive national characteristics.

In all, eight battalions continue to have a NATO reinforcement role, to the ARRC, and the remainder are involved in home defence or 'national defence' as it is now officially termed. This exercise in lateral thinking did much to soften the impact of Options for Change for the TA, and the outcry as far as the Infantry were concerned was muted.

1994 saw an additional infantry battalion deployed to Bosnia. This was the newly amalgamated Royal Gloucestershire, Wiltshire and Berkshire Regiment which joined The Duke of Wellington's Regiment there. While active service is certainly the best way to cement an amalgamation, the Regiment did, however, suffer two early tragedies when four of their members were killed in two separate accidents involving their Saxon wheeled APCs. But, as Autumn, 1994, loomed and hopes that a negotiated peace could be arranged faded, the UN troops were faced with growing uncertainty as it became increasingly likely that the United States would arrange for the arms embargo imposed on the Bosnian Muslims to be lifted. More weapons flooding into this troubled state would merely increase the bloodshed and make the UN role there untenable. Thus plans for what might well be a difficult and dangerous withdrawal began to be drawn up.

Perhaps another and potentially more rewarding role was given in 1994 to a small detachment of the newly created Princess of Wales's Royal Regiment which was sent to Rwanda with the British UN contingent in order to help sort out the chaos resulting from the bloody civil war which had taken place there during the early summer. Even so, increasing fears that the defeated troops of the former government were preparing to launch a guerrilla campaign against the new government also bred uncertainty as to the future.

An even more potentially significant event came on 31 August, 1994, when the IRA finally responded to the December 1993 Major-Reynolds Downing Street Declaration and themselves declared a 'cessation of all military operations'. This was followed a few weeks later by a similar announcement from the Loyalist terrorists. For the British troops on the streets the first result of this was that they were able to remove their helmets and don their berets. Indeed, they welcomed the reduction in tension, but at the same time were aware that the road towards a permanent peaceful resolution to Ulster's problems might still be a long one.

Yet the British Army of the mid-1990s, the Infantry and Royal Armoured Corps in particular, is increasingly conscious of the fact that the era of cuts has not finished. If a UN withdrawal from Bosnia comes about and all sides in Ireland can agree on the future of the North it is probably inevitable that Her Majesty's Treasury will wield its axe once more. Indeed, rumours that the Armed Forces will have to show 'productivity', like any other walk of life, and continuing cost-cutting studies, are already being seen as threats that could still lead to the break-up of the traditional regimental system.

The future of the British Infantry, and especially of its regiments, is an uncertain one. This is largely because of the world situation, especially in eastern Europe, the former Soviet empire, and the powder keg that is the Middle East. Twice, too, during the last decade the totally unexpected has cropped up, the Falklands and the Gulf War, and Yugoslavia could be put under the same category. What is clear, though, is that, whatever the nature of the crisis, infantry will be required. The possible scenarios range from mechanized warfare, as in the Gulf in 1991, airmobile operations, jungle and mountainous warfare, through amphibious operations, to internal security, peacekeeping and, perhaps of growing importance, peacemaking. It means that the infantryman's flexibility, which has grown steadily over the past 330 years, will have to be even greater than ever. This requires ever better training and ever increased motivation. In the past the latter has been generated to a significant degree by the spirit nurtured by the regimental system, and the belief within a regiment that nothing is impossible to achieve.

A number of the arguments used by the detractors of the regimental system have been aired already. Another assertion is that, apart from the

Royal Armoured Corps (together with the Household Cavalry) and the Infantry, the remainder of the British Army does not use it and yet functions perfectly well. This is true, but it is only RAC and Infantry who are traditionally expected to physically close with the enemy on the battlefield; the other arms and services are there to support them in their endeavours to do this. The infantryman, even compared to the tank crewman, is in an especially unique position, as General Sir William Scotter, the then Vice Chief of the General Staff, described to a conference of senior infantry commanders in 1977:

'In the Navy, only the Captain can run away; not so in the Army. And the motivation needed to persuade the individual to do his duty in the face of the enemy is even different in the infantry from that of other Arms; it is different because the infantryman's battlefield environment is not the same as it is for others. For instance, the frightened cavalryman is still close to the rest of his crew and can touch them; and knows he is a vital member of it. Whether he is the driver, the gunner, the radio operator or the tank commander he can clearly visualize that failure on his part can lead to the annihilation of the whole crew. Similarly it is plain to the frightened artilleryman that his individual failure can affect the rest of the crew and the gun. He is further strengthened in his resolve to carry on by that knowledge that those he is supporting are likely to be in a worse case than him. But the frightened infantryman, taking shelter in a shellscrape and out of immediate contact with the rest of his section, can so easily persuade himself that by staying put he alone will have no effect on the course of the battle. What makes him get up and fight is that his mates are doing so and trust him to support them as he trusts them.'

It is this close-knit system which has given the infantryman much of his moral strength over the years. But there is more to it than merely the concept of small group loyalty generated by living and working together. Mutual trust and loyalty must be vertical as well as horizontal. In other words, they must be developed upwards and downwards among the private soldier, his NCOs and his officers. One of the tragedies of Vietnam was that the US policy of rotating personnel through battalions, which were kept permanently stationed there, meant that unit cohesion suffered severely because of the continual turbulence. This, in turn, meant that mutual trust and loyalty were often never given the chance to develop, leading, in some cases, to incidents of 'fragging', when men in combat killed officers and NCOs, either because they did not like them or through fear that they were leading them into danger.

Unit cohesion is also helped by the fact that its members share a common background. One of the most obvious examples of this is that they come from the same part of the country. This has always been the strength of the old County regiments, giving each its own particular character. Likewise, a particular traditional role has helped. The Guardsman, the Paratrooper, Greenjacket and Light Infantryman each have a different approach and their battalions vary widely in character. There is also the tribal instinct. Just as football fans wear the colours of their club to identify them, so the infantry soldier wants his battalion to be easily recognizable. Hence the cap badge and other embellishments are important; the Devon and Dorsets' Croix de Guerre and Royal Welch Fusiliers' flash still mean something to those who wear them today because peculiarities like these set them apart from the rest. Similarly, the soldier wants to belong to a 'good mob', as does the ardent football supporter, who will be at pains to absorb the history of his club. This is why, still today, those joining a regiment for the first time are briefed on its history. Its also gives the infantryman of today identifiable standards to live up to. Finally, there is bonding provided by all ranks growing up together within the same regiment, the essence of the 'regimental family', something which also provides material help for the soldier and his dependants who subsequently fall on hard times after he (and now she) has left the service.

The detractors of the regimental system, and there are a significant number within the Army as well as outside it, tend to be motivated by one of three causes. Either they have never been part of the system, serving with arms and services which do not practice it, and hence are jealous, or they believe that it is archaic and does not rest easily with British society as it approaches the 21st century. Finally, there are those who consider that the Army places an unnecessary administrative yoke around its neck in trying to maintain it, especially when it comes to deploying troops for operations and in restructuring the Army. What most critics appear to overlook is that the infantryman's ultimate task is to be prepared for and to carry out active operations wherever the Government chooses to send him. His ability effectively to perform the tasks set before him, whether they be on the streets of Belfast, the deserts of the Middle East, the hills of Yugoslavia, or wherever, in the face of the whole gamut of weaponry from bricks to nuclear weapons is paramount. He is the point of the military sword, however it is wielded, and all means must be provided to give him the necessary courage, inspiration and moral will to enable him to achieve the aim set before him.

During the past 330 years the British infantryman has faced numerous stiff challenges and there are few parts of the world that he has not trod with his musket or rifle and pack at some time or another. Wherever it has been he has earned the respect of his fellow countrymen, his foes and bystanders. His character, though, remains largely unchanged. While he has never been a saint, he continues to exhibit those enduring qualities of doggedness, courage, loyalty, humour and humanity that he has always done. He has drawn much of his moral strength from 'love of regiment', even though the regimental system has suffered many changes over the centuries and has constantly appeared under threat. Yet, it has always survived and there is no reason why it should not continue to do so, in spite of the anguish caused by inevitable tinkerings with it that will continue to occur. Those who, in the future, find themselves responsible for reshaping the Army to face threats as yet unknown should always bear in mind the words of Francis Bacon, the famous English philosopher of the early 17th century: 'Infantry is the nerve of an army'. Make much of it and cosset it, and the Army as a whole will continue to fulfil the high expectations that the British people, and other nations throughout the world, have always had of it.

The Lineage of the Regiments
(Foot Guards and Line)

REGIMENTS are listed by their year of entering the British Army establishment (English up until 1707). Regimental seniority, apart from the Foot Guards, was not officially recognized until 1694, and then only for the regiments serving in the Low countries. Regimental numbers were not formally allotted to all regiments until 1715, but it was only in 1751 that they officially came into use, although, in practice the designation of regiments by their Colonels' names continued for some years after this.

Significant regimental titles are shown in bold.

For post-amalgamation history see under senior regiment in the amalgamation.

1661

The Royal Regiment of Guards (raised by Charles II when in exile in 1656) – amalgamated with the King's Own Regiment of Foot Guards to become the Royal Regiment of Foot Guards 1665 First Regiment of Foot Guards 1685 First or Grenadier Regiment of Foot Guards 1815 **Grenadier Guards** 1877.
Nicknames The Sandbags, the Coalheavers, The Old Eyes, The Bermuda Exiles, The Bill Browns (3rd Battalion only)

The King's Own Regiment of Foot Guards – amalgamated with The Royal Regiment of Guards 1665.

The Lord General's (Duke of Albemarle's) Regiment of Foot Guards (raised as Colonel Monck's Regiment of Foot from five companies each of the New-castle-based Hesilrige's and Fenwick's Regiments in 1650) – The Coldstream Regiment of or 2nd Foot Guards 1677 **Coldstream Guards** 1855.
Nicknames The Coldstreamers, Nulli Secundus Club

Lord George Douglas's (later Earl of Dumbarton's) Regiment of Foot (**1st**

Foot) (originally raised by Sir John Hepburn in 1633 for service under France) – Retitled The Royal Regiment of Foot 1678 retitled 1st (The Royal Scots) Regiment of Foot 1871 retitled The Lothian Regiment (The Royal Scots) 1881 The Royal Scots (The Lothian Regiment) 1882 **The Royal Scots (The Royal Regiment)** 1920.

Nicknames Pontius Pilate's Bodyguard

1st Tangier Regiment of Foot (**2nd Foot**) – Retitled The Queen's Regiment of Foot 1684 The Queen Dowager's Regiment of Foot 1685 The Queen's Royal Regiment 1703 HRH The Princess of Wales's Own Regiment of Foot 1714 The Queen's Own Royal Regiment of Foot 1727 The 2nd (Queen's Royal) Regiment 1855 The Queen's (Royal West Surrey) Regiment 1881 **The Queen's Royal Regiment (West Surrey)** 1921. Amalgamated with **The East Surrey Regiment (31st/70th Foot)** in 1959 to form **The Queen's Royal Surrey Regiment**. See **The Queen's Regiment** (entered under 1966).

Nicknames The Tangerines, Kirke's Lambs, The Sleepy Queens, The Lambs, The Mutton Lancers.

1662

The Earl Linlithgow's Regiment of Foot Guards (originally raised as Argyle's Regiment in 1639, becoming the Lyfe Guard of Foot 1650, but ceased to exist the following year) – retitled Scotch Guards 1686 Third Regiment of Foot Guards 1711 Scots Fusilier Guards 1831 **Scots Guards** 1877.

Nicknames The Jocks, The Kiddies

1664

The Lord High Admiral's Regiment (Duke of York's Maritime Regiment of Foot) reduced 1684, but brought back to strength and retitled Prince George, Hereditary Prince of Denmark's Regiment of Foot in same year. Disbanded 1689.

1665

The Holland Regiment of Foot (**3rd Foot**) (originally raised in 1572 in the Dutch service) – Retitled Prince George of Denmark's Regiment of Foot 1689 3rd (or The Buffs) Regiment of Foot 1751 3rd (or East Kent, The Buffs) Regiment of Foot 1855 The Buffs (East Kent Regiment) 1881 **The Buffs (Royal East Kent Regiment)** 1935. Amalgamated with **The Queen's Own Royal West Kent Regiment (50th/97th Foot)** in 1960 to

form **The Queen's Own Buffs, The Royal Kent Regiment**. See **The Queen's Regiment** (entered under 1966).
Nicknames The Buff Howards, The Nutcrackers, The Resurrectionists, The Old Buffs, The Admiral's Regiment

1666

Dalzell's Regiment of Foot – raised in Scotland. Disbanded 1667.

1667

The following regiments of Foot were raised, but disbanded after the Treaty of Breda in the same year:

Earl of Chesterfield's
Viscount Townshend's
Lord Alington's
Earl of Ogle's
Marquis of Worcester's
Earl of Manchester's
Bassett's
Apsley's
Sayer's
Vane's
Norton's
Law's

1672

Fitzgerald's Regiment of Foot – raised for sea service, but disbanded 1674.
Duke of Buckingham's Regiment of Foot – disbanded 1673.
Lockhart's Regiment of Foot – raised in Scotland for sea service. Disbanded 1674.
The Marine Regiment – disbanded 1674.

1673

The following regiments of Foot were raised, but disbanded in 1674:

Lord Belasize's
Earl of Ogle's
Earl of Carlisle's
Earl of Peterborough's

Marquis of Worcester's
Duke of Albemarle's
Lord Vaughan's

1674

Monro's Regiment – raised in Scotland. Disbanded 1676.

1678

The following regiments of Foot were raised, but disbanded in 1679:

Duke of Monmouth's
HRH Duchess of York's (Villiers's)
Wheeler's
Lord Alington's
Legg's
Fenwick's
Lord Douglas's
Walden's
Lord Morpeth's
Lord O'Brien's
Sidney's
Goodrick's
Slingsby's
Stradling's

1680

The Earl of Plymouth's or 2nd Tangier Regiment of Foot **(4th Foot)** – retitled The Duchess of York and Albany's Regiment 1684 The Queen Consort's Own Regiment of Foot 1685 The Queen's Marine Regiment 1703 The King's Own Regiment 1715 4th The King's Own Regiment of Foot 1747 The King's Own (Royal Lancaster) Regiment 1881 **The King's Own Royal Regiment (Lancaster)** 1921. Amalgamated with **The Border Regiment(34th/55th Foot)** in 1959 to form **The King's Own Royal Border Regiment**.
Nicknames Barrell's Blues, The Lions

1685

Monk's Regiment (formerly 1st English Regiment Anglo-Dutch Brigade – raised in Holland 1674), but more commonly known as the Irish Regiment

(5th Foot) – retitled 5th Regiment of Foot 1751 The 5th (Northumberland) Regiment of Foot 1782 The Northumberland Fusiliers 1881 **The Royal Northumberland Fusiliers** 1935. See **The Royal Regiment of Fusiliers** (1968).
Nicknames The Shiners, Lord Wellington's Bodyguard, The Fighting Fifth, The Old and Bold, Old Bold 5th

Belasyse's Regiment (formerly 2nd English Regiment Anglo-Dutch Brigade – raised in Holland 1673) **(6th Foot)** – retitled 6th (1st Warwickshire) Regiment of Foot 1782 6th (The Royal 1st Warwickshire Regiment) Foot 1832 The Royal Warwickshire Regiment 1881 **The Royal Warwickshire Fusiliers** 1962. See **The Royal Regiment of Fusiliers** (1968).
Nicknames Guise's Geese, The Warwickshire Lads, The Saucy Sixth

Lord Cutts's Regiment of Foot (raised on Dutch establishment as Disney's in 1674) – disbanded 1698

Mackay's Regiment of Foot (raised as part of Scotch Brigade on Dutch establishment in 1674) – reverted to Dutch establishment 1701 and was still part of it when disbanded in 1782.

Balfour's Regiment of Foot (raised as part of Scotch Brigade on Dutch establishment in 1674) – reverted to Dutch establishment 1701. Disbanded 1717.

Wauchope's Regiment of Foot (raised as part of Scotch Brigade on Dutch establishment in 1674) – reverted to Dutch establishment 1701. Disbanded 1717.

The Royal Regiment of Fuziliers, also The Ordnance Regiment (Lord Dartmouth's) **(7th Foot)** – retitled 7th Regiment of Foot (Royal Fusiliers) 1715, **The Royal Fusiliers (City of London Regiment)** 1881. See **The Royal Regiment of Fusiliers** (1968).
Nicknames The Hanoverian White Horse, The Elegant Extracts

The Princess Anne of Denmark's Regiment of Foot (Lord Ferrers's) **(8th Foot)** – retitled The Queen's Regiment 1702 The King's Regiment of Foot 1716 The 8th (The King's) Regiment of Foot 1751 The King's (Liverpool Regiment) 1881 **The King's Regiment (Liverpool)** 1921. Amalgamated with **The Manchester Regiment (63rd and 96th Foot)** in 1958 to form **The King's Regiment (Manchester and Liverpool)**. Retitled **The King's Regiment** 1968.

Nicknames The Leather Hats, The King's Men

Cornwell's Regiment of Foot **(9th Foot)** – retitled The 9th (East Norfolk) Regiment of Foot 1782 The Norfolk Regiment 1881 **The Royal Norfolk Regiment** 1935. Amalgamated with **The Suffolk Regiment (12th Foot)** in 1959 to form **The 1st East Anglian Regiment (Royal Norfolk and Suffolk)**. See **Royal Anglian Regiment** (1963).
Nicknames The Holy Boys, The Fighting 9th, The Norfolk Howards

Granville's Regiment of Foot (changed to Earl of Bath's) **(10th Foot)** – retitled The 10th (North Lincolnshire) Regiment of Foot 1782 The Lincolnshire Regiment 1881 **The Royal Lincolnshire Regiment** 1946. Amalgamated with **The Northamptonshire Regiment (48th/58th Foot)** in 1960 to form **The 2nd East Anglian Regiment (Duchess of Gloucester's Own Royal Lincolnshire and Northamptonshire)** See **Royal Anglian Regiment** (1963).
Nicknames The Springers, The Poachers

The Duke of Beaufort's Musketeers **(11th Foot)** – retitled The 11th (North Devonshire) Regiment of Foot 1782 **The Devonshire Regiment** 1881. Amalgamated with **The Dorsetshire Regiment (39th/54th Foot)** in 1958 to form **The Devonshire and Dorset Regiment**.
Nicknames The Bloody Eleventh

The Duke of Norfolk's (changed to Earl of Lichfield's) Regiment of Foot **12th Foot)** – retitled The 12th (East Suffolk) Regiment of Foot 1782 **The Suffolk Regiment** 1881. Amalgamated with **The Royal Norfolk Regiment (9th Foot)** in 1959 to form **The 1st East Anglian Regiment (Royal Norfolk and Suffolk)**. See **The Royal Anglian Regiment** (1963).
Nicknames The Old Dozen

The Earl of Huntingdon's Regiment of Foot **(13th Foot)** – retitled The 13th (1st Somersetshire) Regiment of Foot 1782 The 13th (1st Somersetshire Light Infantry) Regiment 1822 The 13th (1st Somersetshire) (Prince Albert's Light Infantry) Regiment 1842 The Prince Albert's (Somerset Light Infantry) 1881 **The Somerset Light Infantry (Prince Albert's)** 1921. Amalgamated with **The Duke of Cornwall's Light Infantry (32nd/46th Foot)** in 1959 to become **The Somerset and Cornwall Light Infantry**. See **The Light Infantry** (1968).
Nicknames Pierce's Dragoons, The Bleeders, The Illustrious Garrison, The Yellow-Banded Robbers

Hales's Regiment of Foot **(14th Foot)** – retitled The 14th (Bedfordshire) Regiment of Foot 1782 The 14th (Buckinghamshire) Regiment of Foot 1809 The 14th (Buckinghamshire – The Prince of Wales's Own) Regiment of Foot 1876 The Prince of Wales's Own (West Yorkshire Regiment) 1881 **The West Yorkshire Regiment (The Prince of Wales's Own)** 1920. Amalgamated with **The East Yorkshire Regiment (15th Foot)** in 1958 to form **The Prince of Wales's Own Regiment of Yorkshire**.
Nicknames The Old and Bold, Calvert's Entire, The Powos

Clifton's Regiment **(15th Foot)** – retitled The 15th (York, East Riding) Regiment of Foot 1782 **The East Yorkshire Regiment (The Duke of York's Own)** 1921. Amalgamated with **The West Yorkshire Regiment (14th Foot)** in 1958 to form the **Prince of Wales's Own Regiment of Yorkshire**.
Nicknames The Snappers, The Poona Guards

1688

Douglas's Regiment of Foot **(16th Foot)** – retitled 16th (Buckinghamshire) Regiment of Foot 1782 16th (Bedfordshire) Regiment of Foot 1809 The Bedfordshire Regiment 1881 **The Bedfordshire and Hertfordshire Regiment** 1919. Amalgamated with **The Essex Regiment (44th/56th Foot)** in 1958 to form **The 3rd East Anglian Regiment**. See **The Royal Anglian Regiment** (1963)
Nicknames The Old Bucks, The Peacemakers, The Featherbeds

Fitzpatrick's Regiment of Foot – disbanded 1689.

Richard's Regiment of Foot **(17th Foot)** – retitled The 17th (Leicestershire) Regiment of Foot 1782 The Leicestershire Regiment 1881 **The Royal Leicestershire Regiment** 1946. Became **4th Battalion The Royal Anglian Regiment** 1963. Disbanded 1970.
Nicknames The Bengal Tigers, The Lilywhites

Gage's Regiment of Foot – Disbanded 1689
Duke of Newcastle's Regiment of Foot – Disbanded 1689
Skelton's Regiment of Foot – Disbanded 1697
Carne's Regiment of Foot – raised in Wales. Disbanded 1689
King's Foot Guards – transferred from Irish establishment. Disbanded 1689.

Forbes's Regiment of Foot **(18th Foot)** (previously raised in 1684 on the Irish Establishment as The Earl of Granard's Regiment) – retitled The

Royal Regiment of Ireland 1695 The 18th (Royal Irish) Regiment of Foot 1751 **The Royal Irish Regiment** 1881. Disbanded 1922.
Nicknames Paddy's Blackguards, The Namurs

Hamilton's Regiment of Foot – transferred from Irish establishment. Disbanded 1689.
McElligott's Regiment of Foot – transferred from Irish establishment. Disbanded 1689.
Viscount Mordaunt's Regiment of Foot – To sea service 1698 disbanded 1699.
Colyear's Regiment of Scots Foot – disbanded 1689?
Earl of Monmouth's Regiment of Foot – disbanded 1689?
Guise's Regiment of Foot – disbanded 1697.

Lutterell's Regiment of Foot **(19th Foot)** – sea service 1702–13 retitled The 19th ((1st Yorkshire North Riding) Regiment of Foot 1782 the 19th (1st Yorkshire North Riding Regiment) (The Princess of Wales's Own) Regiment 1875 Alexandra, Princess of Wales's Own (Yorkshire Regiment) 1881 **The Green Howards (Alexandra, Princess of Wales's Own Yorkshire Regiment)** 1921.
Nicknames The Green Howards (prior to title being made official), The Howard Greens, Howard's Garbage, The Bounders

Peyton's Regiment of Foot **(20th Foot)** – sea service 1702–13 retitled 20th (East Devonshire) Regiment of Foot 1782 **The Lancashire Fusiliers** 1881. See **The Royal Regiment of Fusiliers** (1968).
Nicknames The Two Tens, The Minden Boys, Kingsley's Stand, The Double Xs

1689

O'Farrell's Regiment of Foot **(21st Foot)** (but originally raised in 1678 as the Earl of Mar's Fuziliers in the Dutch service) – retitled The North British Fuziliers 1707 The Royal Regiment of North British Fuziliers 1713 The 21st Regiment of Foot (Royal North British Fuziliers) 1751 21st (Royal Scots Fusiliers) Regiment 1871 **The Royal Scots Fusiliers** 1881. Amalgamated with **The Highland Light Infantry (71st/74th Foot)** in 1959 to form **The Royal Highland Fusiliers (Princess Margaret's Own Glasgow and Ayrshire Regiment)**.
Nicknames The Earl of Mar's Grey Breeks

Duke of Bolton's 1st and 2nd Regiments of Foot – disbanded 1697.

The Duke of Norfolk's Regiment of Foot **(22nd Foot)** – retitled The 22nd (Cheshire) Regiment of Foot 1782 **The Cheshire Regiment** 1881.
Nicknames The Two Twos, The Red Knights, The Lightning Conductors

Lord Herbert's Regiment of Foot **(23rd Foot)** – retitled The Welsh Regiment of Fuziliers 1702 The Royal Regiment of Welsh Fuziliers 1713 The Prince of Wales's Own Royal Regiment of Fuziliers 1714 The Royal Welch Fuziliers 1727 The 23rd (Royal Welch Fuziliers) Regiment of Foot 1751 The Royal Welsh Fusiliers 1881 **The Royal Welch Fusiliers** 1920.
Nicknames The Nanny Goats, The Royal Goats

Dering's Regiment of Foot **(24th Foot)** – retitled The 24th (2nd Warwickshire) Regiment of Foot 1782 **The South Wales Borderers** 1881. Amalgamated with **The Welch Regiment (41st/69th Foot)** in 1969 to form **The Royal Regiment of Wales**.
Nicknames Howard's Greens, The Bengal Tigers

Earl of Drogheda's Regiment of Foot – disbanded 1698.
Viscount Castleton's Regiment of Foot – disbanded 1699.
Earl of Roscommon's Regiment of Foot – disbanded 1698?
Viscount Lisburn's Regiment of Foot – disbanded 1698.
Lord Lovelace's Regiment of Foot – disbanded 1698?
Ingoldsby's Regiment of Foot – disbanded 1698?
Gower's Regiment of Foot – disbanded 1698?
Erle's Regiment of Foot – disbanded 1698
De La Meloniere's Regiment of French Foot – formed from exiled Huguenots. Disbanded 1698.
Du Cambon's Regiment of French Foot – formed from exiled Huguenots. Disbanded 1698.
Marquis La Caillemolte's Regiment of French Foot – formed from exiled Huguenots. Disbanded 1698.
Lord Hamilton's Inniskilling Regiment of Foot – disbanded 1698.
Lloyd's Regiment of Foot – disbanded 1698?
Baker's Londonderry Regiment of Foot – disbanded 1698.
Skeffington's Londonderry Regiment of Foot – disbanded 1698
Reverend George Walker's Londonderry Regiment of Foot – disbanded 1698?

The Earl of Leven's (Edinburgh) Regiment of Foot **(25th Foot)** – retitled The 25th (The Sussex) Regiment of Foot 1782 The 25th (King's Own Borderers) Regiment of Foot 1802 The King's Own Borderers 1881 **The King's Own Scottish Borderers** 1887.

Nicknames The Botherers, The Kokky-Olly Birds, The KOSBs

The Earl of Angus's (Cameronians) Regiment of Foot **(26th Foot)** – retitled 26th (Cameronian) Regiment of Foot 1782. On amalgamation with the **90th Foot** in 1881 became **1st Battalion The Cameronians (Scottish Rifles)**. Suspended animation 1968.

Tiffin's Enniskillen Regiment of Foot **(27th Foot)** – retitled The 27th (Inniskilling) Regiment of Foot 1751. On amalgamation with the **108th Foot** in 1881 became **1st Battalion The Royal Inniskilling Fusiliers.** See **The Royal Irish Rangers** (1968).
Nicknames The Skillingers

Earl of Argyll's Regiment of Foot – disbanded 1697.
Cunningham's Regiment of Foot – disbanded 1697.
Lord Bargeny's Regiment of Foot – disbanded 1689.
Lord Blantyre's Regiment of Foot – disbanded 1689.
Laird of Glencairn's – disbanded 1691.
Viscount Kenmuir's Regiment of Foot – disbanded 1691.
Earl of Mar's Regiment of Foot – disbanded 1689.
Lord Strathnaver's Regiment of Foot – disbanded 1690. Reraised 1693. Disbanded 1717.

1690

1st and 2nd Marine Regiments – amalgamated 1698 disbanded 1699.
Hill's Regiment of Foot – absorbed nine companies of the Laird of Grant's (see 1689 above). Disbanded 1698.

1693

Earl of Donegal's Regiment – disbanded 1698, but see **35th Foot**.
Moncrieff's Regiment – disbanded 1714.

1694

Colonel Robert Mackay's Regiment of Foot – disbanded 1697.

Gibson's Regiment of Foot **(28th Foot)** – disbanded 1697 Reraised 1702 – retitled The 28th (North Gloucestershire) Regiment of Foot 1782. On amalgamation with the **61st Foot** in 1881 became **1st Battalion The Gloucestershire Regiment**. See Options for Change 1992–94.

Nicknames The Old Braggs, The Slashers, The Right-Abouts, The Back Numbers, The Fore and Aft, Brass Before and Brass Behind, The Flowers of Toulouse, The Silver-Tailed Dandies, The Glorious Glosters

Farrington's Regiment of Foot **(29th Foot)** – disbanded 1698 Reraised 1702 – retitled The 29th (Worcestershire) Regiment of Foot 1782. On amalgamation with the **36th Foot** in 1881 became **1st Battalion The Worcestershire Regiment**. Amalgamated with **The Sherwood Foresters (Nottinghamshire and Derbyshire Regiment) (45th/95th Foot)** in 1970 to form **The Worcestershire and Sherwood Foresters Regiment (29th/45th Foot)**.
Nicknames The Ever-Sworded 29th, The Old and Bold, The Firms, The Vein Openers

Northcote's Regiment of Foot – disbanded 1697.
Russell's Regiment of Foot – disbanded 1697?
Lord Murray's Regiment of Foot – disbanded 1697.
Lord Lindsay's Regiment of Foot – disbanded 1698.
Courthorpe's Regiment of Foot – disbanded 1697?
Atkins's Regiment of Foot – disbanded 1697?
Viscount Mountjoy's Regiment of Foot – disbanded 1698. Reraised 1701. Disbanded 1711.
Viscount Charlemont's Regiment of Foot – disbanded 1698, but see also **36th Foot**.
Lillington's Regiment of Foot – disbanded 1697.

1695

Douglas's Regiment of Foot – disbanded 1697.

1698

Four regiments raised, but all disbanded in 1699:

Brudenell's Regiment of Marines
Seymour's Regiment of Marines
Dutton Colt's Regiment of Marines
Brudenell's Regiment of Foot – reraised 1701 sea service 1709 reverted to Foot and disbanded 1713.

1702

Saunderson's 1st Regiment of Marines – became Willis's Regiment of Foot

in 1714 **(30th Foot)** retitled The 30th (1st Cambridgeshire) Regiment of Foot 1782. On amalgamation with the **59th Foot** in 1881 became 1st Battalion The West Lancashire Regiment. Later that year this was changed to **1st Battalion the East Lancashire Regiment.** Amalgamated with **The Prince of Wales's Volunteers (South Lancashire Regiment) (40th/82nd Foot)** in 1958 to form **The Lancashire Regiment (The Prince of Wales's Volunteers).** Amalgamated with **The Loyal Regiment (North Lancashire) (47th/81st Foot)** in 1970 to form **The Queen's Lancashire Regiment.**
Nicknames The Triple XXXs, The Three Tens

Villiers' 2nd Regiment of Marines – became Goring's Regiment of Foot in 1714 **(31st Foot)** retitled 31st (Huntingdonshire) Regiment of Foot 1782. On amalgamation with the **70th Foot** in 1881 became **1st Battalion The East Surrey Regiment.** Amalgamated with **The Queen's Royal Regiment (West Surrey) (2nd Foot)** in 1959 to form **The Queen's Royal West Surrey Regiment.** See **The Queen's Regiment** (1966).
Nicknames The Young Buffs

Fox's 3rd Regiment of Marines – became Borr's Regiment of Foot in 1714 **(32nd Foot)** retitled The 32nd (Cornwall) Regiment of Foot 1782 The 32nd (Cornwall) Light Infantry 1858. On amalgamation with the **46th Foot** in 1881 became **1st Battalion The Duke of Cornwall's Light Infantry.** Amalgamated with **The Somerset Light Infantry (Prince Albert's Own) (13th Foot)** in 1959 to become **The Somerset and Cornwall Light Infantry.** Became **1st Battalion The Light Infantry** 1968.
Nicknames The Docs (DoCs), The Surprisers, The Lacedemonians, The Red Feathers

Viscount Mordaunt's 4th Regiment of Marines – disbanded 1713.
Holt's 5th Regiment of Marines – disbanded 1713.
Viscount Shannon's 6th Regiment of Marines – disbanded 1713.
The Earl of Huntingdon's Regiment **(33rd Foot)** – retitled The 33rd (1st Yorkshire West Riding) Regiment of Foot 1782 The 33rd (Duke of Wellington's Regiment) 1853. On amalgamation with the **76th Foot** in 1881 became **1st Battalion The Duke of Wellington's Regiment (West Riding).**
Nicknames The Havercake Lads

Lord Lucas's Regiment of Foot **(34th Foot)** – sea service 1702–13 retitled The 34th (Cumberland) Regiment of Foot 1782. On amalgamation with the **55th Foot** in 1881 became **1st Battalion The Border Regiment.**

Amalgamated with **The King's Own Royal Regiment (Lancaster) (4th Foot)** in 1959 to form **The King's Own Royal Border Regiment**.
Nicknames The Cattle Reeves

The Earl of Donegal's Regiment of Foot (actually raised in Ireland in 1701) **(35th Foot)** – sea service 1702–13 retitled The 35th (Dorsetshire) Regiment of Foot 1782 The 35th (Sussex) Regiment of Foot 1805 The 35th (Royal Sussex) Regiment 1832. On amalgamation with the **107th Foot** in 1881 became **1st Battalion the Royal Sussex Regiment**. Became **3rd Battalion The Queen's Regiment** in 1966.
Nicknames The Belfast Regiment, The Orange Lilies

Caulfield's Regiment of Foot (actually raised in Ireland in 1701 as Viscount Charlemont's Regiment of Foot) **(36th Foot)** – sea service 1702–13 retitled The 36th (Herefordshire) Regiment of Foot 1782. On amalgamation with the **29th Foot** in 1881 became **2nd Battalion The Worcestershire Regiment**.
Nicknames The Saucy Greens, The Star of the Line, Guards of the Line

Stringer's Regiment of Foot – disbanded 1713.
Lord Mohun's Regiment of Foot – disbanded 1713.
Lord Temple's Regiment of Foot – disbanded 1713.

Meredith's Regiment of Foot **(37th Foot)** – retitled The 37th (North Hampshire) Regiment of Foot 1782. On amalgamation with the **67th Foot** in 1881 became **1st Battalion The Hampshire Regiment**. In 1946 became **The Royal Hampshire Regiment**. See Options for Change 1992–4.
Nicknames The Hampshire Tigers

Lillingstone's Marine Regiment (originally raised for sea service) **(38th Foot)** – became Foot 1705 retitled The 38th (1st Staffordshire) Regiment of Foot in 1782. On amalgamation with the **80th Foot** in 1881 it became **1st Battalion The South Staffordshire Regiment**. Amalgamated with **The Prince of Wales's (North Staffordshire) Regiment** in 1959 to form **The Staffordshire Regiment (Prince of Wales's)**.
Nicknames Pump and Tortoise Brigade

Coote's Regiment of Foot **(39th Foot)** – retitled The 39th (East Middlesex) Regiment of Foot 1782 39th (Dorsetshire) Regiment of Foot 1809. On amalgamation with the **54th Foot** in 1881 became **1st Battalion The**

Dorsetshire Regiment. Amalgamated with **The Devonshire Regiment (11th Foot)** in 1958 to form **The Devonshire and Dorset Regiment**. **Nicknames** Sankey's Horse, The Green Linnets

Grant's Regiment of Scots Foot – disbanded 1713?
Earl of Mar's Regiment of Foot – disbanded 1713.

1703

Allen's Regiment of Foot – disbanded 1713.
Bowles's Regiment of Foot – disbanded 1712.
Evans's Regiment of Foot – disbanded 1713.
Elliot's Regiment of Foot – disbanded 1713.
Pearce's Regiment of Foot – disbanded 1711.

1704

Maccartney's Regiment of Scots Foot – disbanded 1713.
Rooke's Regiment of Foot – disbanded 1712.
Lord Paston's Regiment of Foot – disbanded 1712.
Earl of Daloraine's Regiment of Foot – disbanded 1712.
Lord Scott's Regiment of Foot – disbanded 1713.
Earl of Inchiquin's Regiment of Foot – disbanded 1712.
Viscount Dungannon's Regiment of Foot – disbanded 1713.
Viscount Ikerrin's Regiment of Foot – disbanded 1712.
Marquis de Montandre's Regiment of Foot – disbanded 1713.
Caulfield's Regiment of Foot – disbanded 1712.
Earl of Orrery's Regiment of Foot – disbanded 1713.

1705

Wynne's Regiment of Foot – disbanded 1713.
Breton's Regiment of Foot – disbanded 1712.
Lepell's Regiment of Foot – disbanded 1712.
Soames's Regiment of Foot – disbanded 1713.
Hotham's Regiment of Foot – disbanded 1713.

1706

Lord Kerr's Regiment of Foot – disbanded 1712.
Stanwix's Regiment of Foot – disbanded 1712. Reraised 1715 and given precedence above 34th Foot (because of sea service). Disbanded 1718.

Lord Lovelace's Regiment of Foot – disbanded 1712.
Townshend's Regiment of Foot – disbanded 1712.
Viscount Tunbridge's Regiment of Foot – disbanded 1712.
Bradshaigh's Regiment of Foot – disbanded 1713.
Sybourg's Regiment of Foot – disbanded 1712.
D'Auverqueque's Regiment of French Foot – disbanded 1713.
Price's Regiment of Foot – disbanded 1713.

1708

The following regiments of Foot were raised and disbanded in 1712:

Lord Slane's
Brasier's
Jones's
Carles's – raised in Portugal from Spanish Carlists.

1709

Raised as regiments of Foot and disbanded in 1712:

Dalzell's
Wittewrong's

1715

Dubourgay's Regiment of Foot – given precedence after Stanwix's (see 1706). Disbanded 1718.

Hotham's – given precedence after Dubourgay's and above the 34th. Disbanded 1718.

The following were given precedence immediately after the 39th Foot, but were disbanded in 1718:

Grant's
Pocock's
Lucas's

1716

The following regiments of Foot were formed in Ireland, but were disbanded in 1717:

Trylawny's
Marquis de Montandre's
Creighton's
Wittewrong's
Fielding's
Kane's
Nassau's

1717

Philip's Regiment of Foot **(40th Foot)** – raised from independent companies in Nova Scotia. Retitled the 40th (2nd Somersetshire) Regiment of Foot 1782. On amalgamation with the **82nd Foot** in 1881 became **1st Battalion The Prince of Wales's Volunteers (South Lancashire**. Amalgamated with **The East Lancashire Regiment (30th/59th Foot)** in 1958 to become **The Lancashire Regiment (The Prince of Wales's Volunteers)**. Amalgamated with **The Loyal Regiment (North Lancashire) (47th/81st Foot)** in 1970 to form **The Queen's Lancashire Regiment**.
Nicknames The Excellers (XL), The Fighting Fortieth

1719

Fielding's Regiment of Foot (The Regiment of Invalids) **(41st Foot)** – retitled The Royal Invalids 1747 41st (Royal Invalids) Regiment of Foot 1782 Converted to a full regiment of the Line 1787 41st (The Welsh) Regiment of Foot 1831. On amalgamation with the **69th Foot** in 1881 became **1st Battalion The Welch Regiment**. Amalgamated with **The South Wales Borderers (24th Foot)** in 1969 to form **The Royal Regiment of Wales**.
Nicknames The Invalids, The Old Fogeys

1739

The Earl of Crawford's Regiment of Foot (originally raised as six independent companies of the Highland Regiment during 1725–9) **(42nd Foot)** – retitled The 42nd (Royal Highland) Regiment of Foot 1758 The 42nd Royal Highland Regiment of Foot (The Black Watch) 1861. On amalgamation with the **73rd Foot** in 1881 became 1st Battalion The Black Watch (Royal Highlanders), which was modified in 1936 to **1st Battalion The Black Watch (Royal Highland Regiment)**.
Nicknames The Forty-Twas

The following Marine regiments raised, given Line status as 44th–49th Regiments of Foot in 1741, but transferred to the Admiralty in 1747:

Wolfe's (1st Marine) Regiment
Robinson's (2nd Marine) Regiment
Lowther's (3rd Marine) Regiment
Tyrrell's (4th Marine) Regiment
Douglass's (5th Marine) Regiment
Morton's (6th Marine) Regiment

1740

The following Marine regiments raised, given Line status as 50th–53rd Foot in 1741, but transferred to the Admiralty in 1747:

Cornwallis's (7th Marine) Regiment
Hamner's (8th Marine) Regiment
Powlett's (9th Marine) Regiment
Jefferies's (10th Marine) Regiment

1741

Fowke's Regiment of Foot (43rd Foot) – disbanded 1748

Campbell's Regiment of Foot (54th Foot) – became **43rd Foot** in 1751 retitled The 43rd (Monmouthshire) Regiment of Foot (Light Infantry) 1782 and later The 43rd (Monmouthshire Light Infantry) Regiment of Foot. On amalgamation with the **52nd Foot** in 1881 it became 1st Battalion The Oxfordshire Light Infantry and then, in 1908, **1st Battalion The Oxfordshire and Buckinghamshire Light Infantry**. See **The Green Jackets** (1958).
Nicknames The Ox and Bucks, The Light Bobs, Wolfe's Own

Long's Regiment of Foot (55th Foot) – became **44th Foot** in 1751 retitled The 44th (East Essex) Regiment of Foot 1782. On amalgamation with the **56th Foot** in 1881 became **1st Battalion the Essex Regiment**. Amalgamated with **The Befordshire and Hertfordshire Regiment (16th Foot)** in 1958 to form **The 3rd East Anglian Regiment**. See **The Royal Anglian Regiment** (1963).
Nicknames The Two Fours, The Little Fighting Fours

Houghton's Regiment of Foot (56th Foot) – became **45th Foot** in 1751 retitled The 45th (Nottinghamshire) Regiment of Foot 1779 The 45th

(Nottinghamshire – Sherwood Foresters) Regiment of Foot 1866. On amalgamation with the **95th Foot** in 1881 became **1st Battalion The Sherwood Foresters (Nottinghamshire and Derbyshire Regiment)**. Amalgamated with **The Worcestershire Regiment (29th/36th Foot)** in 1970 to form **The Worcestershire and Sherwood Foresters**.
Nicknames the Old Stubborns, The Sweeps, The Hosiers

Price's Regiment of Foot (57th Foot) – became **46th** Foot in 1751 retitled The 46th (South Devonshire) Regiment of Foot 1782. On amalgamation with the **32nd Foot** in 1881 became **2nd Battalion The Duke of Cornwall's Light Infantry**.
Nicknames Murray's Bucks, The Edinburgh Regiment

Mordaunt's Regiment of Foot (58th Foot) – became **47th Foot** in 1751 retitled The 47th (The Lancashire) Regiment of Foot 1782. On amalgamation with the **81st Foot** in 1881 became 1st Battalion The Loyal North Lancashire Regiment, changed in 1921 to **1st Battalion The Loyal Regiment (North Lancashire)**. Amalgamated with **The Lancashire Regiment (Prince of Wales's Volunteers) (30th/40th/59th/82nd Foot)** in 1970 to form **The Queen's Lancashire Regiment**.
Nicknames The Cauliflowers, The Lancashire Lads, Wolfe's Own

Cholmondeley's Regiment of Foot (59th Foot) – became **48th Foot** in 1751 retitled The 48th (Northamptonshire) Regiment of Foot 1782. On amalgamation with the **58th Foot** in 1881 became **1st Battalion The Northamptonshire Regiment**. Amalgamated with **The Royal Lincolnshire Regiment (10th Foot)** in 1960 to form **The 2nd East Anglian Regiment (Duchess of Gloucester's Own Royal Lincolnshire and Northamptonshire**. See **The Royal Anglian Regiment** (1963).
Nicknames The Heroes of Talavera

Hope's Regiment of Foot (60th Foot) – disbanded 1748

1742

Richbell's Regiment of Foot (61st Foot) – disbanded 1748

1743

Battereau's Regiment of Foot (62nd Foot) – disbanded 1748

Trelawny's Regiment of Foot (63rd Foot) (formed from independent garrison companies in Jamaica known as The Jamaica Volunteers)) – became **49th Foot** in 1751 retitled The 49th (Hertfordshire) Regiment of Foot 1782 The 49th (Princess Charlotte of Wales's or Hertfordshire) Regiment of Foot 1816. On Amalgamation with the **66th Foot** in 1881 became 1st Battalion Princess Charlotte of Wales's (Berkshire Regiment), then 1st Battalion Princess Charlotte of Wales's (Royal Berkshire Regiment) in 1885, and finally, in 1921, **1st Battalion The Royal Berkshire Regiment (Princess Charlotte of Wales's)**. Amalgamated with the **The Wiltshire Regiment (62nd/99th Foot)** in 1959 to form **The Duke of Edinburgh's Royal Regiment (Berkshire and Wiltshire)**. See Options for Change 1992–94.
Nicknames The Biscuit Boys

Earl of Loudon's Regiment of Foot (64th Foot) (Argyll Militia) – disbanded 1747

1745

The following two regiments were raised in North America, disbanded in 1746, reraised in 1754 as the 50th and 51st Foot, but disbanded again in 1756 after being captured at the Battle of Oswego:

Shirley's Regiment of Foot (65th Foot)
Pepperell's Regiment of Foot (66th Foot)

The following regiments were raised, but disbanded in 1746, after the end of the Jacobite Rebellion:

Duke of Bolton's Hampshire Regiment of Foot (67th Foot)
Duke of Bedford's Regiment of Foot (68th Foot)
Montague's Ordnance Regiment (69th Foot)
Earl of Ancaster's Regiment of Foot (70th Foot)
Duke of Rutland's Regiment of Foot (71st Foot)
Berkeley's Regiment of Foot (72nd Foot)
Cholmondeley's Regiment of Foot (73rd Foot)
Duke of Halifax's Regiment of Foot (74th Foot)
Earl of Falmouth's Cornish Regiment of Foot (75th Foot)
Gower's Regiment of Foot (77th Foot)
Herbert's Regiment of Foot (78th Foot)
Edgecombe's Regiment of Foot (79th Foot)

52nd Foot (Abercromby's Regiment) – renumbered **50th Foot** in 1756 retitled The 50th (West Kent) Regiment of Foot 1782 The 50th (The Duke of Clarence's) Regiment of Foot 1827 The 50th Queen's Own Regiment of Foot 1831. On amalgamation with the **97th Foot** in 1881 became 1st Battalion The Royal West Kent Regiment (Queen's Own), and finally in 1922 **1st Battalion The Queen's Own Royal West Kent Regiment**. Amalgamated with **The Buffs (Royal East Kent Regiment) (3rd Foot)** in 1960 to form **The Queen's Own Buffs, The Royal Kent Regiment**. See **The Queen's Regiment** (1966).
Nicknames The Blind Half-Hundred, The Dirty Half-Hundred, The Devil's Royals, The Gallant 50th

53rd Foot (Napier's Regiment) – renumbered **51st Foot** in 1756 retitled The 51st (2nd Yorkshire, West Riding) Regiment of Foot 1782 The 51st (2nd Yorkshire, West Riding) Regiment of Foot (Light Infantry) 1809 The 51st (2nd Yorkshire West Riding) The King's Own Light Infantry Regiment 1821. On amalgamation with the **105th Foot** in 1881 became **1st Battalion The King's Own Yorkshire Light Infantry**. See **The Light Infantry** (1968).
Nicknames The Kolies, The Koylies

54th Foot (Lambton's Regiment) – renumberd **52nd Foot** in 1756 retitled The 52nd (Oxfordshire) Regiment of Foot 1782 The 52nd (Oxfordshire) Regiment of Foot (Light Infantry) 1803. On amalgamation with the **43rd Foot** in 1881 became 2nd Battalion The Oxfordshire Light Infantry and then, in 1909, **2nd Battalion The Oxfordshire and Buckinghamshire Light Infantry**.
Nicknames The Light Bobs

55th Foot (Whitmore's Regiment) – renumbered **53rd Foot** in 1756 retitled The 53rd (Shropshire) Regiment of Foot 1782. On amalgamation with the **85th Foot** in 1881 became **1st Battalion The King's Shropshire Light Infantry**. See **The Light Infantry** (1968).
Nicknames The Brickdusts, The Old Five and Threepennies

56th Foot (Campbell's Regiment) – renumbered **54th Foot** in 1756 retitled the 54th (West Norfolk) Regiment of Foot 1782. On amalgamation with the **39th Foot** in 1881 became **2nd Battalion The Dorsetshire Regiment**.
Nicknames The Two Fives

57th Foot (Perry's Regiment) – renumbered **55th Foot** in 1756 retitled The 55th (Westmorland) Regiment of Foot 1782. On amalgamation with the **34th Foot** in 1881 became **2nd Battalion The Border Regiment**.
Nicknames The Two Fives

58th Foot (Manners' Regiment) – renumbered **56th Foot** in 1756 retitled The 56th (West Essex) Regiment of Foot 1782. On amalgamation with the **44th Foot** in 1881 became **2nd Battalion The Essex Regiment**.
Nicknames The Pompadours, The Saucey Pompeys

59th Foot (Arabin's Regiment) – renumbered **57th Foot** in 1756 retitled The 57th (West Middlesex) Regiment of Foot 1782. On amalgamation with the **77th Foot** in 1881 became 1st Battalion The Duke of Cambridge's Own (Middlesex Regiment) and in 1922 **1st Battalion The Middlesex Regiment (Duke of Cambridge's Own)**. Became **4th Battalion The Queen's Regiment** in 1966. Disbanded 1970.
Nicknames The Steelbacks, The Diehards

60th Foot (Anstruther's Regiment) – renumbered **58th Foot** in 1756 retitled The 58th (Rutlandshire) Regiment of Foot 1782. On amalgamation with the **48th Foot** in 1881 became **2nd Battalion The Northamptonshire Regiment**.
Nicknames the Black Cuffs, The Steelbacks

61st Foot (Montague's Regiment) – renumbered **59th Foot** in 1756 retitled The 59th (2nd Nottinghamshire) Regiment of Foot 1782. On amalgamation with the **30th Foot** in 1881 became 2nd Battalion The West Lancashire Regiment, changed later that year to **2nd Battalion The East Lancashire Regiment**.
Nicknames The Lilywhites

62nd Foot (Campbell's Regiment) in N America as the 62nd Loyal American Provincials – renumbered **60th Foot** in 1756 retitled The 60th (The Royal American) Regiment of Foot 1782 60th King's Royal Rifle Corps Regiment of Foot 1830 and **The King's Royal Rifle Corps** in 1881. See **The Green Jackets** (1958).
Nicknames The Green Jackets, The Jaegers, The 60th Rifles

1756

61st Foot (Forbes Regiment) – renumbered 76th Foot in 1758 disbanded 1763.

2nd Battalions for 3rd, 4th, 8th, 11th, 12th, 19th, 20th, 23rd, 24th, 31st–34th, 36th, 37th Foot raised

1757

62nd Foot (Montgomery's Highlanders) – renumbered 77th Foot in 1758 disbanded 1763.
63rd Foot (Fraser's Highlanders) – renumbered 78th Foot in 1758 disbanded 1767.
64th Foot (Draper's Regiment) – renumbered 79th Foot in 1758 disbanded 1763.

1758

2nd Battalions formed in 1756 renumbered as follows:

2/3rd Foot – **61st Foot** (Elliott's) retitled The 61st (South Gloucestershire) Regiment of Foot 1782. On amalgamation with the **28th Foot** in 1782 became **2nd Battalion The Gloucestershire Regiment**.
Nicknames The White-washers

2/4th Foot – **62nd Foot** (Strode's) retitled The 62nd (Wiltshire) Regiment of Foot 1782. On amalgamation with the **99th Foot** in 1881 became **1st Battalion The Wiltshire Regiment**. Amalgamated with **The Royal Berkshire Regiment (Princess Charlotte of Wales's) (49th/66th Foot)** in 1959 to form **The Duke of Edinburgh's Royal Regiment (Berkshire and Wiltshire)**. See Options for Change 1992–94.
Nicknames The Moonrakers, The Springers, The Splashers

2/8th Foot – **63rd Foot** (Watson's) retitled The 63rd (West Suffolk) Regiment of Foot 1782. On amalgamation with the **96th Foot** in 1881 became **1st Battalion The Manchester Regiment**. Amalgamated with **The King's Regiment (Liverpool) (8th Foot)** in 1958 to form **The King's Regiment (Manchester and Liverpool)**. Retitled **The King's Regiment** 1968.
Nicknames The Bloodsuckers, The Bendovers, The British Musketeers

2/11th Foot – **64th Foot** (Barrington's) retitled The 64th (2nd Staffordshire) Regiment of Foot 1782. On amalgamation with the **98th Foot** in 1881 became 1st Battalion The Prince of Wales's (North Staffordshire Regiment) and in 1921 **1st Battalion The North Staffordshire Regiment (The Prince of Wales's)**. Amalgamated with **The South Staf-**

fordshire Regiment (38th/80th Foot) in 1959 to form The Staffordshire Regiment (Prince of Wales's).
Nicknames The Black Knots

2/12th Foot – 65th Foot (Armiger's) retitled The 65th (2nd Yorkshire, North Riding) Regiment of Foot 1782. On amalgamation with the 84th Foot in 1881 became 1st Battalion The York and Lancaster Regiment. Suspended animation 1968.
Nicknames The Tigers, The Cat and Cabbage, The Twin Roses

2/19th – 66th Foot (Sandford's) retitled The 66th (Berkshire) Regiment of Foot 1782. On amalgamation with the 49th Foot in 1881 became 2nd Battalion Princess Charlotte of Wales's (Berkshire Regiment) in 1885, and finally, in 1921, became 2nd Battalion The Royal Berkshire Regiment (Princess Charlotte of Wales's).
Nicknames The Brave Boys of Berks

2/20th Foot – 67th Foot (Wolfe's) retitled The 67th (South Hampshire) Regiment of Foot 1782. On amalgamation with the 37th Foot in 1881 became 2nd Battalion the Hampshire Regiment and in 1946 2nd Battalion The Royal Hampshire Regiment.
Nicknames Nil

2/23rd Foot – 68th Foot (Lambton's) retitled The 68th (Durham) Regiment of Foot 1782 The 68th (Durham) Light Infantry 1808. On amalgamation with the 106th Foot in 1881 became 1st Battalion The Durham Light Infantry. See The Light Infantry (1968).
Nicknames The Faithful Durhams

2/24th Foot – 69th Foot (Coleville's) retitled The 69th (South Lincolnshire) Regiment of Foot 1782. On amalgamation with the 41st Foot in 1881 became 2nd Battalion The Welch Regiment.
Nicknames The Old Agamemnons, The Ups and Downs

2/31st Foot – 70th Foot (Parslow's) retitled The 70th (Surrey) Regiment of Foot 1782 The 70th (Glasgow Lowland) Regiment of Foot 1813 The 70th (Surrey) Regiment of Foot 1825. On amalgamation with the 31st Foot in 1881 became 2nd Battalion The East Surrey Regiment.
Nicknames The Glasgow Greys

2/32nd Foot – 71st Foot (Petitot's) disbanded 1763.
2/33rd Foot – 72nd Foot (Richmond's) disbanded 1763.

2/34th Foot – 73rd Foot (?) – disbanded 1763.

2/36th Foot – 74th Foot (Talbot's) – disbanded 1763.

2/37th Foot – 75th Foot (Boscawen's) – disbanded 1763.

80th Foot (Gage's) – disbanded 1763

81st Foot (Lindares') (Invalids) – renumbered 71st Foot 1763, but disbanded in same year.

82nd Foot (Parker's) (Invalids) – renumbered 72nd Foot 1763, but disbanded in same year.

83rd Foot (Seabright's) (Invalids) – converted to full regiment of the line 1759 disbanded 1763.

84th Foot (Coote's) (Invalids) – converted to full regiment of the line 1759 disbanded 1763.

1759

85th Foot (Crawfurd's) (Royal Volunteers) – disbanded 1763.

86th Foot (Worge's) – disbanded 1763.

87th Foot (Campbell's Highlanders) – disbanded 1763 reraised 1778 retitled The 87th (Campbell's Highlanders) Regiment of Foot 1782 disbanded 1784.

88th Foot (Keith's Highlanders) – disbanded 1763.

89th Foot (Morris's Highlanders) – disbanded 1765.

90th Foot (Morgan's) – disbanded 1763.

91st Foot (Blayney's) – disbanded 1763.

1760

92nd Foot (Gore's) – disbanded 1763.

93rd Foot (Sutherland's) – disbanded 1763.

94th Foot (Vaughan's) (Royal Welch Volunteers) – disbanded 1763.

95th Foot (Burton's) – disbanded 1763.

96th Foot (Monson's) – disbanded 1763.

97th Foot (Stuart's) – disbanded 1763.

1761

98th Foot – disbanded 1763.

99th Foot – disbanded 1763.

100th Foot – disbanded 1763.

101st Foot – disbanded 1763.

102nd Foot – disbanded 1763.

103rd Foot – disbanded 1763.

104th Foot – disbanded 1763.

105th Foot – disbanded 1763.

106th Foot – disbanded 1763.

107th Foot (The Queen's Royal British Volunteers) – disbanded 1763.

108th Foot – disbanded 1763.

109th Foot – disbanded 1763.

110th Foot – disbanded 1763.

111th Foot – disbanded 1763.

112th Foot – disbanded 1763.

113th Foot – disbanded 1763.

114th Foot – disbanded 1763.

115th Foot – disbanded 1763.

116th Foot (Invalids) – renumbered 73rd Foot 1763, but disbanded in same year.

117th Foot (Invalids) – renumbered 74th Foot 1763, but disbanded in same year.

118th Foot (Invalids) – renumbered 75th Foot 1763, but disbanded in same year.

119th Foot – disbanded 1763.

1762

120th Foot – disbanded 1763.

121st Foot – disbanded 1763.

122nd Foot – disbanded 1763.

123rd Foot – disbanded 1763.

124th Foot – disbanded 1763.

1777

72nd Foot – retitled The 72nd Foot (Royal Manchester Volunteers) 1782 disbanded 1783.

73rd Foot – from an independent company as MacLeod's Highlanders. Renumbered **71st (Highland) Foot** 1786 retitled The 71st (Glasgow Highland) Regiment of Foot 1808 The 71st (Glasgow Highland) Regiment of Foot (Light Infantry) 1809 71st (Highland) Regiment of Foot (Light Infantry) 1810 The 71st (Highland Light Infantry) Regiment of Foot 1855. On amalgamation with the **74th Foot** in 1881 became 1st Battalion The Highland Light Infantry and in 1923 **1st Battalion The Highland Light Infantry (City of Glasgow Regiment)**. Amalgamated with **The Royal Scots Fusiliers (21st Foot)** in 1959 to form **The Royal Highland**

Fusiliers (Princess Margaret's Own Glasgow and Ayrshire Regiment).
Nicknames The Glesca' Kilties

74th Foot – retitled 74th (Argyll Highlanders) Regiment of Foot 1782 disbanded 1783.
76th Foot – retitled the 76th (Macdonald Highlanders) Regiment of Foot 1782 disbanded 1783.
77th Foot – retitled The 77th (Atholl Highlanders) Regiment of Foot 1782 disbanded 1783.

78th Foot (Seaforth's Highlanders) – renumbered **72nd (Highland) Regiment of Foot** in 1786 retitled the 72nd (The Duke of Albany's Own Highlanders) Regiment of Foot 1823. On amalgamation with the **78th Foot** in 1881 became **1st Battalion The Seaforth Highlanders (Ross-shire Buffs, The Duke of Albany's)**. Amalgamated with **The Queen's Own Cameron Highlanders (79th Foot)** in 1960 to form **The Queen's Own Highlanders (Cameron and Seaforth)**. See Options for Change 1992–94.
Nicknames The Wild Macraes

80th Foot – retitled The 80th (Royal Edinburgh Volunteers) Regiment of Foot 1782 disbanded 1784.
83rd Foot – retitled The 83rd (Royal Glasgow Volunteers) Regiment of Foot 1782 disbanded 1783.
84th Foot – retitled The 84th (Royal Highland Emigrants) Regiment of Foot 1782 disbanded 1784.

1778

75th Foot – retitled The 75th (Prince of Wales's) Regiment of Foot 1782 disbanded 1783.
79th Foot – retitled The 79th (Liverpool) Regiment of Foot 1782 disbanded 1789.
85th Foot – retitled The 85th (Westminster Volunteers) Regiment of Foot 1782 disbanded 1783.

1779

86th Foot – retitled The 86th (Rutland Volunteers) Regiment of Foot 1782 disbanded 1784.
88th Foot – disbanded 1784.

89th Foot – disbanded 1784.

90th Foot – retitled The 90th (Yorkshire Volunteers) Regiment of Foot 1782 disbanded 1784.

91st Foot – retitled The 91st (Shropshire Volunteers) Regiment of Foot 1782 disbanded 1784.

92nd Foot – disbanded 1783.

1780

93rd Foot – disbanded 1783.

94th Foot – disbanded 1783.

95th Foot – disbanded 1784.

96th Foot – disbanded 1784.

97th Foot – disbanded 1784.

98th Foot – disbanded 1783.

99th Foot (Jamaica Regiment) – disbanded 1783.

100th Foot – disbanded 1783.

1781

101st Foot – disbanded 1785.

102nd Foot – disbanded 1784.

103rd Foot – disbanded 1784.

1782

104th Foot – raised from independent companies in N America (formed 1778) and disbanded 1784.

105th Foot – raised from independent companies in N America (formed 1778). Transferred to Irish establishment 1783 disbanded 1787.

1787

2/42nd Foot becomes **The 73rd (Highland) Regiment of Foot** – retitled The 73rd (Perthshire) Regiment of Foot 1862. On amalgamation with the **42nd Foot** in 1881 became 2nd Battalion The Black Watch (Royal Highlanders) and then, in 1936, **2nd Battalion The Black Watch (Royal Highland Regiment)**.

Nicknames Nil

74th (Highland Regiment of Foot – retitled The 74th (Highlanders) Regiment of Foot 1847. On amalgamation with the **71st Foot** in 1881

became 2nd Battalion The Highland Light Infantry and then in 1923 **2nd Battalion The Highland Light Infantry (City of Glasgow Regiment)**.
Nicknames The Assayes, The Pig and Whistle Regiment

75th (Highland) Regiment of Foot – retitled 75th (Stirlingshire) Regiment of Foot 1862. On amalgamation with the **92nd Foot** in 1881 became **1st Battalion The Gordon Highlanders**. See Options for Change 1992–94.
Nicknames The Gay Gordons

76th Foot – raised at expense of East India Company and also known, until 1812, as the Hindoostan Regiment. On amalgamation with the **33rd Foot** in 1881 became 2nd Battalion The Duke of Wellington's (West Riding Regiment) and in 1921 **2nd Battalion The Duke of Wellington's Regiment (West Riding)**.
Nicknames The Immortals, The Pigs, The Seven-and-Sixpennies

77th Foot – retitled The 77th (East Middlesex) Regiment of Foot 1807, and The 77th (East Middlesex) Regiment (Duke of Cambridge's Own) 1876. On amalgamation with the **57th Foot** in 1881 became 2nd Battalion The Duke of Cambridge's Own (Middlesex Regiment), changed in 1922 to **2nd Battalion The Middlesex Regiment (Duke of Cambridge's Own)**.
Nicknames The Pot-Hooks

1793

78th (Highland) Regiment of Foot – 2nd Battalion (The Ross-shire Buffs) raised 1794 two battalions amalgamated as the 78th (Highland) Regiment of Foot (Ross-shire Buffs) 1796. On amalgamation with the **72nd Foot** in 1881 became **2nd Battalion The Seaforth Highlanders (Ross-shire Buffs, The Duke of Albany's)**.
Nicknames The King's Men

79th Foot (The Cameronian Volunteers) – retitled the 79th Regiment of Foot (Cameronian Highlanders) 1804 The 79th Regiment of Foot (Cameron Highlanders) 1806 The 79th Queen's Own Cameron Highlanders 1873, and finally in 1881 **The Queen's Own Cameron Highlanders**. Amalgamated with **The Seaforth Highlanders (Ross-shire Buffs, The Duke of Albany's (72nd/78th Foot)** in 1960 to form **The Queen's Own Highlanders (Seaforth and Cameron)**. See Options for Change 1992–4.

Nicknames Nil

80th Foot (Staffordshire Volunteers) – on amalgamation with the **38th Foot** in 1881 became **2nd Battalion The South Staffordshire Regiment**.
Nicknames The Staffordshire Knots

81st Foot (Loyal Lincoln Volunteers) – on amalgamation with the **47th Foot** in 1881 became 2nd Battalion The Loyal North Lancashire Regiment and then, in 1921, **2nd Battalion The Loyal Regiment (North Lancashire)**.
Nicknames Nil

82nd Foot (Prince of Wales's Volunteers) – on amalgamation with the **40th Foot** in 1881 became 2nd Battalion The Prince of Wales's Volunteers (South Lancashire) and then, in 1938, **2nd Battalion The South Lancashire Regiment (The Prince of Wales's Volunteers)**.
Nicknames Nil

83rd Foot (The County of Dublin Regiment) – on amalgamation with the **86th Foot** in 1881 became 1st Battalion The Royal Irish Rifles, changed to **1st Battalion The Royal Ulster Rifles** in 1922. See **The Royal Irish Rangers** (1968).
Nicknames Fitch's Grenadiers

84th Foot – retitled The 84th (York and Lancaster) Regiment of Foot 1809. On amalgamation with the **65th Foot** in 1881 became **2nd Battalion The York and Lancaster Regiment**.
Nicknames Nil

85th Foot (Buck's Volunteers) – retitled The 85th King's Light Infantry 1815. On amalgamation with the **53rd Foot** in 1881 became **2nd Battalion The King's Shropshire Light Infantry**.
Nicknames The Elegant Extracts, The Young Bucks

86th Foot (Shropshire Volunteers) – retitled The 86th (Leinster) Regiment of Foot 1809 The 86th (Royal County of Down) Regiment of Foot 1812. On amalgamation with the **83rd Foot** in 1881 became 2nd Battalion The Royal Irish Rifles, changed to **2nd Battalion The Royal Ulster Rifles** in 1921.
Nicknames The Irish Giants

87th Foot (The Prince of Wales's Irish) – retitled The 87th (Prince of Wales's Own Irish) Regiment of Foot 1811 The 87th (Prince of Wales's Own Irish Fusiliers) Regiment of Foot 1827, and later in the same year The 87th (Royal Irish Fusiliers) Regiment of Foot. On amalgamation with the **89th Foot** in 1881 became 1st Battalion Princess Victoria's (Royal Irish Fusiliers), which was altered in 1921 to **1st Battalion The Royal Irish Fusiliers (Princess Victoria's)**. See **The Royal Irish Rangers (1968)**.
Nicknames The Aiglers, The Eagle Takers, The Rollickers, The Old Fogs, The Faugh-a-Ballagh Boys

88th Foot (Connaught Rangers) Amalgamated with the 94th Foot in 1881 to become **1st Battalion The Connaught Rangers**. Disbanded 1922.
Nicknames The Devil's Own

89th Foot – retitled The 89th (Princess Victoria's) Regiment of Foot 1866. On amalgamation with the **87th Foot** in 1881 became **2nd Battalion Princess Victoria's (Royal Irish Fusiliers)**.
Nicknames Blayney's Bloodhounds

93rd Foot – disbanded 1797.
95th Foot – disbanded 1796.
96th Foot – disbanded 1796.
102nd Foot – disbanded 1795.
103rd Foot – disbanded 1795.
116th (Perthshire) Foot – disbanded 1795.
131st Foot – disbanded 1796.

1794

90th Foot (Perthshire Volunteers) – retitled The 90th Light Infantry (Perthshire Volunteers) 1815. On amalgamation with the **26th Foot** in 1881 became **2nd Battalion The Cameronians (Scottish Rifles)**.
Nicknames The Perthshire Grey Breeks

91st Foot – disbanded 1795.
92nd Foot – disbanded 1795.
94th Foot – disbanded 1796.
97th (Inverness-shire) Foot – disbanded 1796.

98th (Argyllshire) Regiment of Foot (Highlanders) – renumbered **91st Foot** 1798 retitled The 91st (Argyllshire) Regiment of Foot 1820 The 91st

(Argyllshire Highlanders) Regiment of Foot 1864 The 91st (Princess Louise's Argyllshire Highlanders) Regiment of Foot 1872. On amalgamation with the **93rd Foot** in 1881 became 1st Battalion Princess Louise's (Sutherland and Argyll Highlanders), changed to Princess Louise's (Argyll and Sutherland Highlanders) in 1882. In 1920, in a final title change, it became **1st Battalion The Argyll and Sutherland Highlanders (Princess Louise's)**. Reduced to a cadre, The Balaklava Company, in 1971 but restored to full battalion size in 1972.
Nickname The Rories

99th Foot – disbanded 1797.

100th Foot (Gordon Highlanders) – renumbered **92nd Foot** 1798 retitled The 92nd (Gordon Highlanders) Regiment of Foot 1861. On amalgamation with the **75th Foot** in 1881 became **2nd Battalion The Gordon Highlanders**.
Nicknames The Gay Gordons

101st Foot – disbanded 1795.
104th Foot (Royal Manchester Volunteers) – disbanded 1796.
105th Foot – disbanded 1795.
106th (Bulwer's) Foot – disbanded 1795.
107th (Keating's) Foot – disbanded 1795.
108th Foot – disbanded 1795.
109th (Aberdeenshire) Foot – disbanded 1795.
110th Foot – disbanded 1795.
111th (Loyal Birmingham Volunteers) Foot – disbanded 1795.
112th Foot – disbanded 1795.
113th Foot – disbanded 1795.
114th Foot – disbanded 1795.
115th (Prince William's) Foot – disbanded 1795.
117th Foot – disbanded 1796.
118th Foot – disbanded 1795.
119th Foot – disbanded 1796.
120th Foot – disbanded 1796.
121st (County Clare) Foot – disbanded 1796
122nd (Londonderry) Foot – disbanded 1796.
123rd (Leatherband's) Foot – disbanded 1796.
124th (Beresford's) Foot – disbanded 1795.
125th Foot – disbanded 1796.
126th Foot – disbanded 1796.
127th Foot – disbanded 1796.

128th Foot – disbanded 1796.
129th (Gentlemen of Coventry) Foot – disbanded 1796.
130th Foot – disbanded 1796.
132nd (Highland) Foot – disbanded 1796.
133rd (Highland) Foot – disbanded 1796
134th (Royal Limerick) Foot – disbanded 1796.
135th Foot – disbanded 1796.

1800

93rd (Highland) Regiment of Foot (formed from The Sutherland Fencibles) – retitled The 93rd (Sutherland Highlanders) Regiment of Foot 1861. On amalgamation with the **91st Foot** in 1881 became 2nd Battalion Princess Louise's (Sutherland and Argyll Highlanders), changed in 1882 to Princess Louise's (Argyll and Sutherland Highlanders). In 1920 became **2nd Battalion The Argyll and Sutherland Highlanders (Princess Louise's)**. **Nicknames** The Thin Red Line

1802

94th Foot (Scotch Brigade) – disbanded 1817.

95th Rifle Regiment – became The Rifle Brigade in 1816 and was removed from the register of line regiments. Retitled The Prince Consort's Own Rifle Brigade 1862 The Rifle Brigade (Prince Consort's Own) 1868 The Prince Consort's Own (Rifle Brigade) 1882 and, finally, in 1920 **The Rifle Brigade (Prince Consort's Own)**. See **The Green Jackets** (1958). **Nicknames** The Sweeps, The Greenjackets

96th Foot – renumbered 95th Foot in 1816. Disbanded 1817.

1803

97th (Queen's Own Germans) Foot (British Musketeers) – raised from the Minorca Regiment, composed of Germans and Swiss. Renumbered 96th Foot 1816 disbanded 1817.

1804

98th Foot – renumbered 97th Foot in 1816 disbanded 1817.
99th Foot – retitled The Prince of Wales's Tipperary 1811 renumbered 98th Foot in 1816 disbanded 1817.

100th Foot – retitled HRH The Prince Regent's County of Dublin Regiment 1812, renumbered 99th Foot in 1816 disbanded 1817.

1805

101st Foot (Duke of York's Royal Irish) – renumbered 100th Foot in 1816 disbanded 1817

1808

102nd Foot – raised from the New South Wales Corps (Australia) renumbered 101st Foot 1816 disbanded 1817.
103rd Foot – raised from 103rd Garrison Battalion. Renumbered 102nd foot 1816 disbanded 1817.

1810

104th Foot – raised from The New Brunswick Fencibles. Renumbered 103rd Foot 1816 disbanded 1817.

1823

94th Foot – Inherited some of the battle honours and traditions of the previous 94th (see 1802). Amalgamated with **88th Foot** in 1881 to become **2nd Battalion The Connaught Rangers.**
Nicknames The Garvies

95th (Derbyshire) Foot – Amalgamated with **45th Foot** in 1881 to become **2nd Battalion The Sherwood Foresters (Nottinghamshire and Derbyshire Regiment).**
Nicknames The Nails

1824

96th Foot – Inherited some of the previous battle honours and traditions of the previous 96th (see 1802), Amalgamated with **63rd Foot** in 1881 to become **2nd Battalion The Manchester Regiment.**
Nicknames The Bendovers

97th Foot – retitled 97th (Earl of Ulster's) Regiment of Foot 1826. Amalgamated with **50th Foot** in 1881 to become 2nd Battalion The Queen's Own (Royal West Kent Regiment), changed to The Royal West

Kent Regiment (Queen's Own) in 1921, and finally to **2nd Battalion The Queen's Own Royal West Kent Regiment** in 1922.
Nicknames The Celestials

98th Foot – retitled The 98th (Prince of Wales's) Regiment of Foot 1876. On amalgamation with **64th Foot** in 1881 became 2nd Battalion The Prince of Wales's (North Staffordshire Regiment), changed in 1921 to **2nd Battalion The North Staffordshire Regiment**
Nicknames Nil

99th (Lanarkshire) Foot – retitled The 99th (Duke of Edinburgh's) Regiment of Foot 1876. Amalgamated with **62nd Foot** in 1881 to become 2nd Battalion The Duke of Edinburgh's (Wiltshire Regiment), changed in 1920 to **2nd Battalion The Wiltshire Regiment (Duke of Edinburgh's)**.
Nicknames Nil

1858

100th Foot (The Prince of Wales's Royal Canadian Regiment) – Amalgamated with **109th Foot** in 1881 to become 1st Battalion The Prince of Wales's Royal Canadian Regiment and later that year **1st Battalion The Prince of Wales's Leinster Regiment (Royal Canadians)**. Disbanded 1922
Nicknames The Beavers, The Colonials, The Centipedes

The following were transferred from the Honourable East India Company:

1st Bengal Fusiliers (originally raised as The Bengal European Regiment (Honourable East India Company) in 1756) – retitled **The 101st (Royal Bengal Fusiliers) Regiment of Foot** 1862. Amalgamated with the **104th Foot** in 1881 to become **1st Battalion The Royal Munster Fusiliers**. Disbanded 1922.
Nicknames The Dirty Shirts.

1st Madras Fusiliers (originally raised as The Honourable East India Company's European Regiment in 1748) – retitled **The 102nd (Royal Madras Fusiliers) Regiment of Foot** 1862. Amalgamated with the **103rd Foot** in 1881 to become **1st Battalion The Royal Dublin Fusiliers**. Disbanded 1922.
Nicknames The Blue Caps

1st Bombay Fusiliers (originally raised as The Bombay Regiment in 1661) – retitled **The 103rd (Royal Bombay Fusiliers) Regiment of Foot** 1862. Amalgamated with the 102nd Foot in 1881 to become **2nd Battalion The Royal Dublin Fusiliers**. Disbanded 1922.
Nicknames The Old Toughs

2nd Bengal Fusiliers (originally raised as the 2nd Bengal (European) Regiment (Honourable East India Company) in 1839) – retitled **The 104th (Bengal Fusiliers) Regiment of Foot** 1862. Amalgamated with the 101st Foot in 1881 to become **2nd Battalion The Royal Munster Fusiliers**. Disbanded 1922.
Nickname The Dirty Shirts

2nd Madras Light Infantry (originally raised as The 2nd Madras (European Light Infantry) Regiment (Honourable East India Company) in 1839) – retitled **The 105th (Madras Light Infantry) Regiment of Foot** 1862. Amalgamated with the 51st Foot in 1881 to become **2nd Battalion The King's Own Yorkshire Light Infantry**.
Nicknames Nil

2nd Bombay Light Infantry (originally raised as The 2nd Bombay European Regiment (Honourable East India Company) in 1826) – retitled **The 106th (Bombay Light Infantry) Regiment of Foot** 1862. Amalgamated with the 68th Foot in 1881 to become **2nd Battalion The Durham Light Infantry**.
Nicknames Nil

3rd Bengal Light Infantry (originally raised as the 3rd Bengal European Light Infantry (Honourable East India Company in 1854) – retitled **The 107th (Bengal Infantry) Regiment of Foot)** 1862. Amalgamated with the 35th Foot in 1881 to become **2nd Battalion The Royal Sussex Regiment**.
Nicknames Nil

3rd Madras Light Infantry (Honourable East India Company) (formed in 1854) – retitled **The 108th (Madras Infantry) Regiment of Foot** 1862. Amalgamated with the 27th Foot in 1881 to become **2nd Battalion The Royal Inniskilling Fusiliers**.
Nicknames Nil

3rd Bombay Light Infantry (originally raised as the 3rd Bombay European Regiment (Honourable East India Company) in 1854) – retitled **The**

109th (Bengal Infantry) Regiment of Foot 1862. Amalgamated with the **100th Foot** in 1881 to become 2nd Battalion The Prince of Wales's (Royal Canadian) Regiment, later **2nd Battalion The Prince of Wales's Leinster Regiment (Royal Canadian). Disbanded 1922.**
Nicknames The Poona Pets, The Brass Heads

1900

Irish Guards
Nicknames Bob's Own, The Micks

1915

Welsh Guards
Nicknames The Taffs

1942

The Lowland Regiment – disbanded 1949
The Highland Regiment – disbanded 1949

1948

The following were transferred from the Indian Army:

2nd King Edward VI's Own Gurkha Rifles (The Sirmoor Rifles)

6th Gurkha Rifles – retitled **6th Queen Elizabeth's Own Gurkha Rifles** 1959.

7th Gurkha Rifles – retitled **7th Duke of Edinburgh's Own Gurkha Rifles** 1959.

10th Gurkha Rifles – retitled **10th Princess Mary's Own Gurkha Rifles** 1949.

1958

Formation of **The Green Jackets** (retitled **The Royal Green Jackets** 1966) comprising:

1st Battalion (The Oxfordshire and Buckinghamshire Light Infantry (**43rd/52nd Foot)**

2nd Battalion (The King's Royal Rifle Corps) (**60th Rifles**)
3rd Battalion (The Rifle Brigade (Prince Consort's Own))

1963

Formation of **The Royal Anglian Regiment** comprising:

1st Battalion (The 1st East Anglian Regiment) (**9th/12th Foot**)
2nd Battalion (The 2nd East Anglian Regiment) (**10th/48th/58th Foot**)
3rd Battalion (The 3rd East Anglian Regiment) (**16th/44th/56th Foot**)
4th Battalion (The 4th East Anglian Regiment) (**17th Foot**) – disbanded 1970.

1966

Formation of **The Queen's Regiment** comprising:

1st Battalion (The Queen's Royal Surrey Regiment) (**2nd/31st/70th Foot**)
2nd Battalion (The Queen's Own Buffs, Royal Kent Regiment) (**3rd/50th/ 97th Foot**)
3rd Battalion (The Royal Sussex Regiment) (**35th/107th Foot**)
4th Battalion (The Middlesex Regiment (Duke of Cambridge's Own) (**57th/77th Foot**) – disbanded 1970. New 4th Battalion was raised in 1971, but disbanded in 1972.

See Options for Change 1992–94

1968

Formation of **The Royal Regiment of Fusiliers** comprising:

1st Battalion (The Royal Northumberland Fusiliers) (**5th Foot**)
2nd Battalion (The Royal Warwickshire Fusiliers) (**6th Foot**)
3rd Battalion (The Royal Fusiliers) (**7th Foot**)
4th Battalion (The Lancashire Fusiliers) (**20th Foot**) – disbanded 1969

Formation of **The Light Infantry** comprising:

1st Battalion (The Somerset and Cornwall Light Infantry) (**13th/32nd/ 46th Foot**)
2nd Battalion (The King's Own Yorkshire Light Infantry) (**51st/105th Foot**)
3rd Battalion (The King's Shropshire Light Infantry) (**53rd/85th Foot**)

4th Battalion (The Durham Light Infantry) (**68th/106th Foot**) – disbanded 1969

Formation of **The Royal Irish Rangers (27th Inniskilling), 83rd and 87th)** comprising:

1st Battalion (The Royal Inniskilling Fusiliers) (**27th/108th Foot**)
2nd Battalion (The Royal Ulster Rifles) (**83rd/86th Foot**)
3rd Battalion (The Royal Irish Fusiliers (Princess Victoria's)) (**87th/89th Foot**) – disbanded 1968

See Options for Change 1992–94

OPTIONS FOR CHANGE (1992–1994)

The following lost their second battalions:

Grenadier Guards
Coldstream Guards
Scots Guards

The following large regiments lost their third battalions:

The Light Infantry
The Royal Green Jackets
The Royal Regiment of Fusiliers
The Royal Anglian Regiment

The Queen's Regiment merged with **The Royal Hampshire Regiment** to produce a regiment of just two Regular battalions **The Princess of Wales's Royal Regiment (Queen's and Royal Hampshires)** in 1992.

The Royal Irish Rangers merged with **The Ulster Defence Regiment**, forming **The Royal Irish Regiment** with one Regular battalion in 1992.

The Gloucestershire Regiment amalgamated with **The Duke of Edinburgh's Royal Regiment** to form **The Royal Gloucestershire, Wiltshire and Berkshire Regiment** in 1994.

The four regiments of Gurkha Rifles were merged in 1994 to form **The Royal Gurkha Rifles** comprising:

1st Battalion – formerly 2nd and 6th Gurkha Rifles
2nd Battalion – formerly 7th Gurkha Rifles

3rd Battalion – formerly 10th Gurkha Rifles

(One battalion is scheduled for disbandment after Hong Kong is handed over to China in 1997)

The Queen's Own Highlanders (Cameron and Seaforth) amalgamated with **The Gordon Highlanders** in 1994 to form **The Highlanders (Seaforth, Gordons and Camerons)**.

Regiments of Foot Guards and Line
1994

Grenadier Guards
Coldstream Guards
Scots Guards
Irish Guards
Welsh Guards

The Royal Scots (The Royal Regiment) [1st Foot]
The Princess of Wales's Royal Regiment (Queen's and Royal Hampshires) [2nd, 3rd, 31st, 35th, 37th, 50th, 57th, 67th, 70th, 77th, 97th, 107th Foot]
The King's Own Royal Border Regiment [4th, 34th, 55th Foot]
The Royal Regiment of Fusiliers [5th, 6th, 7th, 20th Foot]
The King's Regiment [8th, 63rd, 96th Foot]
The Royal Anglian Regiment [9th, 10th, 12th, 16th, 17th, 44th, 48th, 56th, 58th Foot]
The Devonshire and Dorset Regiment [11th, 39th, 54th Foot]
The Light Infantry [13th, 32nd, 46th, 51st, 53rd, 68th, 85th, 105th, 106th Foot]
The Prince of Wales's Own Regiment of Yorkshire [14th, 15th Foot]
The Green Howards (Alexandra, Princess of Wales's Own Yorkshire Regiment) [19th Foot]
The Royal Highland Fusiliers (Princess Margaret's Own Glasgow and Ayrshire Regiment) [21st, 71st, 74th Foot]
The Cheshire Regiment [22nd Foot]
The Royal Welch Fusiliers [23rd Foot]
The Royal Regiment of Wales [24th, 41st, 69th Foot]
The King's Own Scottish Borderers [25th Foot]
The Royal Irish Regiment [27th, 83rd, 86th, 87th, 89th, 108th Foot]
The Royal Gloucestershire, Wiltshire and Berkshire Regiment [28th, 49th, 61st, 62nd, 66th, 99th Foot]

The Worcestershire and Sherwood Foresters Regiment [29th, 36th, 45th, 95th Foot]

The Queen's Lancashire Regiment [30th, 40th, 47th, 59th, 81st, 82nd Foot]

The Duke of Wellington's Regiment (West Riding) [33rd, 76th Foot]

The Staffordshire Regiment (Prince of Wales's) [38th, 64th, 80th, 98th Foot]

The Black Watch (Royal Highland Regiment) [42nd, 73rd Foot]

The Royal Green Jackets [43rd, 52nd, 60th Foot, Rifle Brigade]

The Highlanders (Seaforth, Gordons and Camerons) [72nd, 75th, 78th, 79th, 92nd]

The Argyll and Sutherland Highlanders (Princess Louise's) [91st, 93rd Foot]

The Parachute Regiment

The Royal Gurkha Rifles

Bibliographical Note

To produce a full list of all sources used for the two volumes of this work would result in a very lengthy bibliography, which would probably do the reader no favours, especially since it would reflect some forty years' worth of research which I have carried out in the realm of military history. Nevertheless, apart from such obvious sources as regimental and campaign histories, and personal accounts, there have been a number of works which have been especially useful.

No history of this genre can be written without referral to Sir John Fortescue's classic multi-volume *History of the British Army* (Macmillan 1879–1930), which still endures on account of the depth of its research. Other useful general sources are Field Marshal Lord Carver's *The Seven Ages of the British Army* (Weidenfeld & Nicolson 1984), Correlli Barnett's *Britain and Her Army* (Allen Lane 1970), and David Ascoli's *Companion to the British Army 1660–1983* (Harrap 1983). A good ready reference is Henry William and Catherine Patricia Adams *History of the British Regular Army 1660–1990: Vol 1 – A Dative Record of the Sovereign's Regiments* (Major Book Publications 1990). As for the nature of the British soldier over the ages, two short books, J. M. Brereton's *The British Soldier: A Social History from 1661 to the Present Day* (Bodley Head 1986) and Victor Neuburg's *Gone for a Soldier: A History of Life in the British Ranks from 1642* (Cassell 1989) provide a wealth of information. As for the British officer, there is no better nor more entertaining account than E. S. Turner's *Gallant Gentlemen: A Portrait of the British Officer 1600–1956* (Michael Joseph 1956).

For a concise and informative treatment, Frederick Myatt's *The British Infantry 1660–1945* (Blandford Press 1983) is much recommended. Charles Dalton's multi-volume *English Army Lists and Commission Registers 1661–1714*, which was published in the 1890s, was of immeasurable help in navigating a course through the tangle that the regiments endured during that period. Thereafter the *Army List* proved an excellent source. David Chandler's *The Art of Warfare in the Age of Marlborough* (Batsford 1976)

and Michael Glover's *Wellington's Army* (David and Charles 1977) are particularly good, as are Hew Strachan's *From Waterloo to Balaclava: Tactics, Technology and the British Army 1815–1854* (Cambridge University Press 1985) and *Wellington's Legacy: The Reform of the British Army 1830–54* (Manchester University Press 1984). The nature of the British Infantry on the eve of the Great War is examined by John Baynes in *Morale: A Study of Men and Courage* (Cassell 1967). Curiously, there is no comprehensive history of the British Army during the Great War, although *A Nation in Arms* (Tom Donovan 1985), edited by Ian Beckett and Keith Simpson, goes a long way towards defining the character of the Army during 1914–18. This is unlike for the Second World War, which is covered by General Sir David Fraser's *And We Shall Shock Them* (Hodder and Stoughton 1983).

A good background to the Army's operations in the 20th century other than the two World Wars is Lawrence James's *Imperial Rearguard: Wars of Empire, 1919–85* (Brassey's 1988). Dwelling specifically on the post–1945 period are Gregory Blaxland's *The Regiments Depart* (Kimber 1971) and the more concise *Brush Fire Wars* (Robert Hale 1984) by Michael Dewar. Finally, there have been a number of studies of the British Army today published during the past decade, but by far the most perceptive of these is Antony Beevor's *Inside the British Army* (Chatto & Windus 1990).

General Index

1. Campaigns, operations, and battles are indicated by dates in brackets.
2. Ranks of individuals are those eventually achieved (at the time of writing) and where known.

Baku, 26
Ballater, 171
Ballymena, 175, 190
Bantams, 16
Bardia, 66
Barracks, 47, 157, 198
Bartholomew Committee (1940), 61–62, 63
Baskeyfield, Sgt John (S Staffs), 121
Basra, 24, 217
Bates, Cpl Sidney (R Norfolks), 121
Bathsheba, 30
Battalion organization, 16, 42, 43, 47–48, 73fn, 81, 112, 113
Battle schools, 63–64
BBC, 124, 143
Beak, Gen Daniel, 93, 94
Beda Fomm (1941), 66
Belfast, 194, 196
Belize (formerly British Honduras), 201
Bennion, Pte (Border R), 141–142
Berbera, 66
Bergen-Belsen, 127
Berlin, 59, 128
Berlin Airlift (1948–49), 167
Bermuda, 192
Berthelot, Gen, 20
Bigland, Alfred, 16
Bir Hacheim (1942), 72, 73
Black and Tans, 37, 38, 40
Blacker, Gen Sir Cecil, 183
Blay, Pte (Queens), 100
Bloody Sunday (1972), 197
Bluff Cove, 203
Boer War (1899–1902), 23
Bois du Buttes (1918), 20
Bologna, 134
Bombay, 77, 149–150
Borneo (1963–66), 179–181
Bosnia (1992–), 221–222, 223
Bou Aoukaz (1943), 97–98
Boulogne, 11
Bradley, Jim, 77
Brand, RQMS (Post Office Rifles), 19
Brandon, Lt Col R. L. (RF), 97
Brecon Beacons, 202

Bredin, Brig 'Bala', 58, 159
Bremen (1945), 128
Breskens, 123
Bridge of Don (Aberdeen), 189
Brigade Training Groups, 156
Briggs, Gen Sir Harold, 160
British Guiana (1953), 170
British Somaliland, 65–66, 67, 68
Brodie, Gen, 139
Brooke, Rupert, 27
Brown, Pte (DCLI), 122
Browning, Gen 'Boy', 123
Brunei Revolt (1962), 179
Bunbury, Lt Col Ramsay (DWR), 167
Burbury, Lt Col Richard (S Lancs), 116
Burma (1941–45), 78–79, 105–110, 134–145
Burma (post WW2), 147–148
Burma-Siam railway, 78
Burnett-Stuart, Gen Jock, 42–43
Bury St Edmunds, 175
Buss, Drummer (Glosters), 165

Caen Canal, 114
Cain, Maj Robert (S Staffs), 121
Cairo, 75, 168
Calais (1940), 56–58, 65
Calcutta, 105, 136, 148, 149
Calistan, Sgt Charles (RB), 83–84
Calvert, Brig Mike, 108, 139
Cameroons (1961), 178
Campbell, Brig Jock, 69
Canal Zone (1951–56), 167–169, 170
Canea, 68
CANLOAN, 122
Canterbury, 175
Cape Town, 77
Caporetto (1917), 32
Cardwell system, 156
Carlin, Sgt (R Innis F), 29
Carne, Brig James, 164–165
Carrington, Charles (R Warwicks), 6
Carver, FM Lord, 181, 182
Caterham, 175
Catterick, 45
Cauldron, The (1942), 73

Wavell, FM Earl, 69, 108, 109
Weapons – anti-tank, 48, 51, 54, 81,
  167, 199, 210; grenades, 16;
  machine guns, 16, 43, 48, 178, 210;
  mortars, 1, 16–17, 47, 48, 81, 99,
  178; rifles, 50, 81, 177–178, 181,
  210; submachine guns, 52–53, 91,
  177
Wehr, 124
Werewolves, 154
Wesel, 126–127
West Wall, 123
Westkapelle, 123
Wheeler, Gen Roger, 220
Wilberforce, Lt Col (E Surreys), 97
Willetts, Sgt Michael (Para), 197
Wilson, Brig Tony, 203
Wilson, Capt Eric (East Surreys), 66
Wilson, FM Lord, 67

Wilson, FM Sir Henry, 39
Winchester, 175, 190
Wingate, Gen Orde, 108, 109, 140,
  138–139
Wireless Ridge (1982), 207
Witzig, Maj Rudolf, 87
Woodall, Rev Hugh (Suffolks), 118
Wormhoudt, 58–59
Wrangel, Gen, 37

Xanten, 126

Yalta Conference (1944), 153
Yalu River, 163
Ypres (1914–18), 1, 5, 6, 11, 12, 17,
  18, 19
Yunnan, 138

Zanzibar (1964), 183

Formation and Regimental Index

9th SS Panzer Div (Hohenstauffen), 120
10th SS Panzer Div (Frundsberg), 121
15th Panzer Div, 94, 127

INDIA

3rd (Special Force) Div, 138
4th Div, 65, 66, 68, 70, 94, 95, 98, 131
5th Div, 68, 73, 134, 136
7th Div, 134, 143
10th Div, 105
11th Div, 77
14th Div, 106–107, 110
17th Div, 78, 79, 135–136
19th Div, 143
20th Div, 141, 143, 146
34th Div, 110
12 Inf Bde, 78
14 Inf Bde, 139–140, 141
16 Inf Bde, 138, 139
18 Inf Bde, 81
47 Inf Bde, 106–107
71 Inf Bde, 107
77 Inf Bde, 108–109, 138, 139–140, 141
111 Inf Bde, 138, 139–140, 141
123 Inf Bde, 106
161 Inf Bde, 136, 137
Assam Rifles, 136
Punjab Regiment, 66, 137
Rajputana Rifles, 131

JAPAN

18th Div, 138
55th Div, 107, 134
56th Div, 138

NEPAL

Shere Regiment, 136

NEW ZEALAND

2nd Div, 92
5 Inf Bde, 82

POLAND

Para Bde, 121

SOUTH AFRICA

1st Div, 82

USA

1st Marine Div, 214
3rd Inf Div, 129
17th Abn Div, 126
82nd Abn Div, 114, 121
101st Abn Div, 114, 121
5th Cav Regt, 166
339th Inf Regt, 36

WEST AFRICA

3 Bde, 139–140